Concepts of materials science

From reviews after publication

I strongly recommend this book as essential reading for anyone considering applying to study materials at university.
Peter Haynes, Head of Department of Materials, Imperial College London

Sutton, materials science's leading pedagogue, has distilled the essence of materials science to its most basic, beautiful ideas and presents it in a manner accessible to every science or engineering student. This is a MUST READ for every young scientist, as well as their teachers!
David J Srolovitz, Dean of Engineering, The University of Hong Kong

This book is a remarkable attempt to provide the background needed to engage with materials science in the 21st century.
Mike Ashby, University of Cambridge

The book will ... be useful at many levels of education. It may steer high school students who have taken advanced physics and chemistry classes toward a college major they probably have never heard of. Likewise, it may excite physics and chemistry majors in college about contributing to this highly interdisciplinary field in graduate school. It will be useful supplementary reading in a college science course for non-scientists. For the professionals, it will be a treat to savor the broad sweep of Sutton's vision of their field.
Frans Spaepen, Harvard University

This book tries successfully to guide students to big ideas and key concepts, without stepping into details. This made this book immensely readable and attractive.
Takeshi Egami, University of Tennessee and Oak Ridge National Laboratory

This book provides a backbone on which readers can build further and deeper studies on specific topics without losing sight of the underlying physical concepts that link together the vast number of topics that materials science includes today.
Frank Ernst, Case Western Reserve University

Much more than an introduction to materials science, ... it is written with great insight, explained in straightforward language and with minimal use of equations. This book ... would be useful supplementary reading for undergraduates looking for the bigger picture on selected topics. I would recommend this book to pre-university students ... and to postgraduates in neighbouring disciplines.
Steve Sheard, University of Oxford

Adrian Sutton's ... exceptional clarity of the text enables ... students to gain a unique insight into the subject in just 132 pages. This book is priceless!
Xanthippi Markenscoff, University of California San Diego

Having taught materials science for so many years, I am gratified to see the clarity of Sutton's treatment, but I am also a bit jealous of his expository abilities.
David P Pope, University of Pennsylvania

Adrian Sutton's direct and informal writing style and his emphasis on physical understanding ... make this marvellous book ... a pleasure to read.
Mojmír Šob, Masaryk University, Czech Republic

Concepts of materials science

Adrian P Sutton FRS

OXFORD
UNIVERSITY PRESS

OXFORD
UNIVERSITY PRESS

Great Clarendon Street, Oxford, OX2 6DP,
United Kingdom

Oxford University Press is a department of the University of Oxford.
It furthers the University's objective of excellence in research, scholarship,
and education by publishing worldwide. Oxford is a registered trade mark of
Oxford University Press in the UK and in certain other countries

Published in the United States of America by Oxford University Press
198 Madison Avenue, New York, NY 10016, United States of America

British Library Cataloguing in Publication Data
Data available

Library of Congress Control Number: 2021934825

ISBN 978–0–19–284683–9 (hbk.)
ISBN 978–0–19–284644–0 (pbk.)

DOI: 10.1093/oso/ 9780192846839.001.0001

Printed and bound by
CPI Group (UK) Ltd, Croydon, CR0 4YY.

In memory of my parents Peter and Beryl Sutton

About the author

Educated in materials science at the Universities of Oxford and Pennsylvania, Adrian Sutton has held professorships at Oxford University, Aalto University and Imperial College London. A materials physicist, addressing fundamental aspects of the science of materials, he has taught across the undergraduate curriculum. He has consulted for companies in the United States, Japan and the UK. He was elected to a Fellowship of the Royal Society in 2003. He is a founder of the Thomas Young Centre, the London Centre for Theory and Simulation of Materials. He was also the founding director in 2009 of the Centre for Doctoral Training (CDT) on Theory and Simulation of Materials at Imperial College, which attracted into materials science more than a hundred graduates from the UK and overseas with first class honours degrees or equivalent in physics and engineering. His book *Rethinking the PhD* is the extraordinary story of this acclaimed CDT. In 2012 he was awarded the Rector's Medal for Outstanding Innovation in Teaching at Imperial College. In 2018 he left paid employment to focus on scholarship, and Imperial College conferred the title of professor emeritus on him. He lives in Oxford with his wife Pat White.

Also by Adrian P Sutton:

Electronic structure of materials

Interfaces in crystalline materials, with R W Balluffi

Rethinking the PhD

Physics of elasticity and crystal defects

Preface

All technologies depend on the availability of suitable materials. Without innovations in materials we would still be living in caves. The epochs of civilisation have been defined by the materials people have used, from the stone age to the bronze age to the iron age and now the silicon age. Mark Miodownik has written eloquently about how countless materials shape our lives[1]. Robert Cahn wrote an authoritative history of the emergence of materials science[2].

This book is different. It is an attempt to identify key concepts – big ideas – of the science of materials. It is not about any particular experimental, theoretical or computational technique or any particular materials. It describes ten concepts I believe are central to materials science.

One difficulty I faced in writing this book was the overlap between materials science and condensed matter physics, solid state chemistry, mechanics of solids and biology. Materials science has become so broad that for a long time I wondered whether there are any concepts that tie the subject together.

I think of materials as a subset of condensed matter, distinguished by having a use in an existing or intended technology. The link to technology is the *raison d'être* of the subject. It explains why the study of materials involves scientists and engineers, and why it is often described as an 'enabling discipline' because it facilitates advances in technology. I have tried in this book to highlight the fundamental nature of materials science and the inexhaustible richness of its intellectual content.

The engineering of materials may be described[3] as the *exploitation* of the relationships between the structure, properties and method of fabrication of a material, to design a material with optimum performance for particular applications. The science of materials may be described as *understanding* those relationships. Understanding and exploiting those relationships to design and create materials for applications in technology is the essence of 'Materials' as a discipline.

To identify core concepts of materials science I had to peel away layers of detail in the search for ideas that pervade the subject. Inevitably some of the concepts are common to mainstream physical sciences, such as thermodynamic stability (Chapters 1 and 2), symmetry (Chapter 5) and quantum behaviour (Chapter 6). In those cases I have focused on their particular significance in materials science.

Thermodynamics defines the stable state of a material in its environment. It is rare for a material to be in this stable state, but it is the state towards which the

[1] Miodownik, M, *Stuff Matters*, Penguin Group (2013).

[2] Cahn, R W, *The Coming of Materials Science*, Elsevier (2001)

[3] National Research Council 1989. *Materials Science and Engineering for the 1990s: Maintaining Competitiveness in the Age of Materials*, Chapter 1: What is materials science and engineering? p.19-34. National Academies Press: Washington DC. https://doi.org/10.17226/758

material will evolve if it is left alone in its environment. This immediately introduces the idea of change in a material, either towards a thermodynamically stable state or some other state determined by its exposure to forces of various kinds. In crystalline materials defects of various kinds are the agents of change (Chapter 4). The rate of change is determined by restless atomic motion in materials, both in facilitating the motion of defects and in retarding them (Chapter 3). Defects and their interactions in crystalline materials are a perfect illustration of the concept of emergence of new physics in materials at larger length scales through collective behaviour at smaller length scales (Chapter 8). The emergence of new physics across the range of length scales from electrons to engineering components is a defining and unique feature of materials science. The ability to manipulate the structure of materials across this range of length scales to achieve desired properties leads to the concept of materials design for particular applications (Chapter 9).

Size matters in materials because their properties are more obviously dominated by quantum physics at the nanoscale (Chapter 7). This has led to the rise of nanoscience and nanotechnology which have underpinned the modern age of information storage and processing. Until metamaterials were introduced around the turn of the 21st century there were no materials that displayed certain properties, such as negative refraction. Metamaterials removed this limitation because their properties are not determined by their chemistry but by their carefully designed structure (Chapter 10). Treating biological matter as a material has led to the concept of active matter in which complexity and self-organisation arise from the collective action of energy-consuming agents (Chapter 11).

I have strived to make this book intelligible for anyone with a pre-university education in physics, chemistry and mathematics. On the whole I have limited the use of mathematics to elementary algebra and quoting the occasional useful formula. Only in Chapter 10 have I relaxed a little this self-imposed discipline. Such a short book as this cannot be self-contained. There are references to books in the further reading at the end of each chapter. References to research papers with hyperlinks in the electronic version are included in footnotes for the reader who wants to delve deeper.

Undergraduates and masters students of materials science may find this book a refreshing and enlightening supplement to their usual reading. Graduates in other subjects may gain an impression of what materials science is about, and I hope it will draw them into postgraduate study of the subject. I hope my colleagues in materials science will find it stimulating and occasionally provocative.

This is *not* a text-book. Much of what is in text-books on materials science is not treated in this book and *vice versa*. Nevertheless, this book does cover a lot of ground.

I am grateful to Bob Balluffi, Craig Carter, Martin Castell, Peter Dobson, Mike Finnis, Peter Haynes, Peter Hirsch, Stan Lynch, Tony Paxton, John Pendry, Bob Pond, Luca Reali, Chris Race, Tchavdar Todorov, Vasek Vitek and anonymous reviewers for helpful comments. Any remaining errors are my responsibility.

Finally, I thank Pat for her support, encouragement and editorial skill.

Imperial College London
December 2020.

Contents

1   **When is a material stable?**   1
   1.1   Concept   1
   1.2   Introduction   1
   1.3   Definitions   2
   1.4   The first law of thermodynamics   4
   1.5   The second law of thermodynamics   5
   1.6   Closed systems and heat reservoirs   12
   1.7   The Helmholtz free energy   13
   1.8   The Gibbs free energy   14
   1.9   Chemical potentials   15
   1.10 The Gibbs-Duhem equation   16
   1.11 The Gibbs phase rule   17
   1.12 Closing remarks   18
   Further reading   18

2   **Phase diagrams**   19
   2.1   Introduction   19
   2.2   Free energy - composition curves   21
   2.3   From free energy - composition curves to the equilibrium state   22
   2.4   Phase diagram for complete miscibility   25
   2.5   Phase diagrams for limited solubility in the solid state 26
   2.6   Closing remarks   28
   Further reading   29

3   **Restless motion**   30
   3.1   Concept   30
   3.2   Evidence of restless atomic motion   30
   3.3   Fluctuations and thermally activated processes   31
   3.4   Brownian motion   33
   3.5   The fluctuation-dissipation theorem   34
   3.6   Some other manifestations of restless atomic motion in materials   37
   Further reading   39

4   **Defects**   40
   4.1   Concept   40
   4.2   Change in materials   40
   4.3   Point defects   41
   4.4   Dislocations   46
   4.5   Grain boundaries   50
   Further reading   51

5   Symmetry   52
   5.1   Concept   52
   5.2   Introduction   52
   5.3   Conservation laws   54
   5.4   Physical properties of crystals   55
   5.5   Topological defects   57
   5.6   Quasicrystals   58
   Further reading   64

6   Quantum behaviour   65
   6.1   Concept   65
   6.2   The size and identity of atoms   65
   6.3   The double slit experiment   66
   6.4   Identical particles, the Pauli exclusion principle and spin   71
   6.5   Consequences of the Pauli exclusion principle   72
   6.6   Tunnelling   76
   6.7   Thermal properties of solids   76
   6.8   Quantum diffusion   79
   6.9   Closing remarks   80
   Further reading   80

7   Small is different   81
   7.1   Concept   81
   7.2   Introduction   81
   7.3   Quantum dots   83
   7.4   Catalysis   86
   7.5   Giant magnetoresistance   86
   7.6   Closing remarks   93
   Further reading   93

8   Collective behaviour   94
   8.1   Concept   94
   8.2   More is different   94
   8.3   Three examples of processes involving multiple length scales   96
   Further reading   101

9   Materials by design   102
   9.1   Concept   102
   9.2   Introduction   102
   9.3   Microstructure   103
   9.4   An example: replacing the 'nickel'   105
   9.5   Self-assembly   105
   9.6   Smart materials   111
   9.7   Closing remarks   113
   Further reading   113

10   Metamaterials   114
   10.1   Concept   114

10.2	Introduction	114
10.3	An example: a metamaterial for elastic waves	115
10.4	Electromagnetic metamaterials and negative refraction	118
10.5	Invisibility cloaks	122
10.6	Closing remarks	123
	Further reading	124
11	**Biological matter as a material**	**125**
11.1	Concept	125
11.2	What is life?	125
11.3	Active matter	127
11.4	Synthetic biology	131
11.5	Closing remarks	131
	Further reading	132
Index		**133**

1
When is a material stable?

A theory is the more impressive the greater the simplicity of its premises, the more different the kinds of things it relates, and the more extended its area of applicability. Hence the deep impression that classical thermodynamics made upon me. It is the only physical theory of universal content concerning which I am convinced that, within the framework of applicability of its basic concepts, it will never be overthrown.

Albert Einstein, Autobiographical Notes p.31, translated and edited by Paul Arthur Schilpp, Open Court Publishing Company (1996). ISBN 0812691792. © The Hebrew University of Jerusalem. With permission of the Albert Einstein Archives and of Cricket Media.

1.1 Concept

When they are not being deformed, irradiated, or constantly disturbed by other means, materials evolve towards a state of equilibrium with their environments. Thermodynamics defines the conditions for such equilibrium under various environmental conditions. Phase diagrams[1] are maps of the equilibrium states of a material as the temperature and concentrations of its constituents are varied, usually at atmospheric pressure.

1.2 Introduction

Most materials are unstable or metastable. A material is in a metastable state if its energy is at a local minimum, but not the lowest energy state. A corrie lake is a metastable location of water. Water in the lake attains its lowest energy when it reaches the sea. If it is to reach the sea without evaporating the water level in the lake has to rise above the lip retaining it. When materials are undisturbed they evolve towards a stable state in whatever environment they are in. The evolution may involve obvious changes taking place such as corrosion, where the surface of the material reacts chemically with gases or liquids in the environment. Less obvious are the changes that may be taking place inside the material. They may also be chemical in nature, involving the movement of atoms to where the changes are taking place. There may be other changes to the internal structure of the material, involving a reduction in the concentration of defects, which I will discuss in Chapter 4. Depending on the temperature of the material and

---

[1]Phase diagrams are discussed in the next chapter.

the energy barriers that have to be overcome, some of these processes may be occurring imperceptibly slowly, in some cases on a geological time-scale (hundreds of millions of years). Others may be occurring in picoseconds (10^{-12} s). The possible range of time-scales for these processes spans 27 orders of magnitude[2].

There are other occasions when materials are driven away from a stable state. For example the zirconium cladding of the nuclear fuel rods in a pressurised water nuclear reactor is being constantly bombarded by neutrons. Low density polythene undergoes chemical changes when it is exposed to sunlight, becoming brittle and releasing methane and ethane. Rails deform and occasionally develop cracks through the repeated loading due to trains passing over them.

Thermodynamics defines the conditions for different regions within a material to be in equilibrium with each other and with any environment the material is in. If the material is able to exchange energy and/or matter with its environment, equilibrium involves those exchanges as well. When a material is in equilibrium it is stable in the sense that it does not undergo any further changes. In this chapter we will find out what those conditions are.

Anyone who studies a physical science or engineering will eventually meet thermodynamics because its concepts and principles are so universal, as Einstein remarked in the quote at the beginning of this chapter. We are going to go on quite a journey in this chapter and I am not going to assume you know anything about thermodynamics. But first we need to define certain terms we will be using.

1.3 Definitions

In thermodynamics the object we are considering and its environment is called the *system*. There are *isolated systems* in which an object is surrounded by boundaries that are impenetrable to energy and matter. The object is then separated from its environment and the system comprises the object only. If the object can exchange energy but not matter with its environment the system is called *closed*. There are also *open systems* in which an object can exchange energy and matter with its environment. Although the overall chemical composition of an object in either an isolated or closed system does not change, there can be a redistribution of elements within the object as it evolves towards equilibrium. The state of equilibrium of an object depends not only on the object itself but also on whether and how it interacts with its environment.

The chemical species that make up the system are called *components*. They may be atoms, such as iron and carbon, or molecules, such as water and methane. The *chemical composition* of a multi-component system is the specification of the concentration of each component present. In a multi-component system it is quite common to find regions in which the atomic structure and chemical composition are constant. Such a region is called a *phase*. If the region occupied by the phase is only nanometres in size then a significant fraction of the atoms will be close to the surface or interface bounding it. The structure and composition of such a small region may then differ significantly from a macroscopic region. It is doubtful that such a small region can be classified as

---

[2]An order of magnitude is a factor of ten. Twenty seven orders of magnitude means a factor of 10^{27}.

a phase. This is one reason why thermodynamics applies only to macroscopic systems, containing large numbers of particles.

We are used to using temperature scales like the Celsius scale[3], the zero of which corresponds to the freezing point of pure water, and for which negative temperatures are possible. In thermodynamics we use an *absolute* temperature scale, measured in kelvin, the zero of which represents an absolute minimum temperature. One way this scale can be defined is with the notion of an *ideal gas*: a collection of perfect point particles which do not interact with one another except when they collide. At a fixed pressure, a quantity of an ideal gas has a volume that is proportional to the temperature measured on the absolute scale, becoming zero at zero temperature. This absolute zero on the Kelvin scale is equal to -273.15 degrees celsius. Therefore, the temperature in kelvin equals the temperature in degrees celsius plus 273.15.

Heat is the *kinetic energy* of the random movement of atoms that make up an object. When hot and cold bodies are brought into contact kinetic energy of atoms in the hot body is transferred to atoms in the cold body through atomic collisions. We say that they have reached the same temperature when there is no further net transfer of atomic kinetic energy between them. At a microscopic level there are local transfers of kinetic energy between both bodies through atomic collisions, but they sum to zero on average when the bodies have the same temperature.

Atoms that make up the object of study also have *potential energy*. It arises from the attractive and repulsive forces they exert on each other, and from their interaction with electric, magnetic and gravitational fields imposed on the object by its environment. In a solid the potential and kinetic energies of each atom are changing continuously and extremely rapidly because each atom is vibrating about its average position. The period of vibration is around 10^{-13} seconds. In a cubic centimetre of a solid there are around 10^{22} atoms. To define the state of a cubic centimetre of a solid in terms of the instantaneous kinetic and potential energies of each atom is impossible. It is also unnecessary. In thermodynamics the state of a system comprising a single component can be defined in terms of only two variables, called *state variables*. For example, the thermodynamic state of a fixed amount of a single component substance is determined by its equation of state, which relates the three state variables pressure, volume and temperature. When any two of these three state variables are specified the third is determined by the equation of state. Furthermore, unlike the kinetic and potential energies of individual atoms these state variables are experimentally measurable. State variables and properties are either extensive or intensive. *Extensive variables* are proportional to the size of the system, such as volume, the amount of a component and internal energy. *Intensive variables* are independent of the size of the system, such as temperature, pressure and chemical potentials (chemical potentials are discussed in section 1.9).

When a material is in equilibrium the intensive variables temperature, pressure and chemical potentials are constant throughout the system. This is the definition of the *equilibrium state of a material* in the absence of any fields acting on the material, such as gravity. When such fields exist they have to be taken into account in the determination of the equilibrium state. For example, the pressure within a column

---

[3]In 1948 the centigrade temperature scale was renamed the celsius scale. They are identical scales.

supporting a tall building has to increase as we go down the column to maintain mechanical equilibrium.

As we have already noted it is common for a material not to be in equilibrium. Provided it is not being driven away from equilibrium this state remains important because it is the state towards which the system evolves. In that case it provides the *direction* of change within the material. But it does not provide the *rate* of change because time does not appear in equilibrium thermodynamics. Although many of the most useful materials are not at equilibrium they are in metastable states that can endure for much longer than the service life of the material. This feature of materials is exploited extensively in their design, as discussed in Chapter 9.

1.4   The first law of thermodynamics

The first law of thermodynamics is the conservation of energy:

Energy cannot be created or destroyed. It can only be converted from one form to another.

As we shall see in Chapter 5 the ultimate origin of this law is a symmetry of a particular kind. Forms of energy include kinetic energy, heat and various forms of potential energy such as chemical energy, electrical energy, magnetic energy, gravitational energy, and so on.

Energy is the capacity to do work[4]. In thermodynamics *work* does not have the usual meaning of labour. It has the same meaning as in mechanics. Work is done when the point of application of a force moves in the direction of the force. The work done is equal to the force multiplied by the displacement of the point of its application in the direction of the force. When work is done energy is transferred. If the force is exerted on an object by its surroundings, and the point of its application is displaced in the direction of the force, energy is transferred from the surroundings to the object, and *vice versa*. When you stretch a spring by pulling on its ends you are doing work on the spring. The work you have done is converted into potential energy in the spring. If you release one end of the spring it retracts quickly and its potential energy is converted into kinetic energy. We shall see that in thermodynamics there are other forms of work, not only those of a mechanical origin.

The *internal energy* of an object can increase in two ways in a closed system. It can receive heat and it can have work done on it. If it loses heat or if it does work its internal energy decreases. In an open system, where the object can exchange matter and energy with its environment, its internal energy can also change through the addition or removal of atoms. If an atom of a particular element is added to the system, with no simultaneous transfer of heat or work, the internal energy increases by an amount called the *chemical potential* of the element. Chemical potentials are intensive variables like temperature and pressure and we shall see that they play a central role in the equilibrium of isolated, closed and open multi-component systems.

---

[4]In the next section we will see there is a limitation on the conversion of heat into work. A more precise statement is that *free* energy is the capacity to do work under specified conditions. Free energy is introduced in section 1.7.

The equilibrium state of a system is not determined by minimisation of the internal energy alone, except at extremely low temperatures. For example, when a solid melts its internal energy increases because it absorbs latent heat, but melting is a change of phase to a new equilibrium state. The other essential ingredient is the *entropy* of the system. In classical thermodynamics entropy is defined somewhat abstractly by the properties of heat engines, which we will not go into here. In the next section we will introduce entropy in more physical terms.

1.5   The second law of thermodynamics

1.5.1   Irreversibility and entropy production

A man is filmed as he falls ten metres into a tall, thermally insulated tank containing 100 cubic metres of water. There is a splash as his body displaces water. No water escapes because the tank walls are high. Waves are created and reflected off the walls. He finally comes to rest afloat, the waves die away leaving a flat surface of the water, and all the water splashes run down the sides of the tank back into the body of water.

We know immediately if the film is played backwards because it shows an impossible sequence of events. How can a man, starting from a stationary position in perfectly still water in a tank, emerge completely dry and fly through the air at a sufficient speed to reach his original position ten metres above the water? Although this is obviously impossible it does not violate the first law of thermodynamics! The potential energy the man had when he was ten metres above the water is converted into his kinetic energy as he falls towards the water and then into kinetic energy of water molecules in the tank. In other words, the water is warmer. According to the first law of thermodynamics there is no reason why the heat that has been imparted to the water cannot be converted back to potential energy by sending him back to his original position above the tank.

By running the film backwards we are seeing a world in which the direction of time has been reversed. The equations governing the man's motion and the motion of the water molecules are exactly the same if time is reversed. They display 'time-reversal symmetry'. A return to his original position above the tank does not violate these equations of motion because they are independent of the direction of time. But all our instincts tell us this is impossible.

What are we missing? In 100 cubic metres of water there are around 10^{30} molecules. Each of these molecules can be anywhere in the 100 cubic metre volume of water. They have a range of velocities, determined by the temperature of the water. Each molecule followed a particular trajectory when the man jumped into the tank. Its trajectory is defined by its position and velocity throughout a time interval, say from just before he reached the surface of the water to some time after the water in the tank has settled down around him. To reverse his fall each water molecule would have to follow the same trajectory but in reverse. That is possible but it is just one of very many trajectories each molecule can follow. The total number of possible trajectories of all 10^{30} molecules is a huge number, but it is not infinite. If time were reversed, it would not violate the conservation of energy if all water molecules reversed their trajectories resulting in the man's ejection from the water. But for that to happen the water molecules would have to follow one particular set of 10^{30} molecular trajectories from a total number of possible trajectories much larger than 10^{30}. The probability of that happening is so small as effectively to be impossible[5].

What we are homing in on here is the *irreversibility* of spontaneous or natural processes. Once we have stirred fresh milk into a cup of tea we find it impossible to separate them. If we run hot and cold water into a bath we cannot separate the hot water from the cold. When we deform a piece of metal, such as a paperclip, so that it takes on a new permanent shape the metal becomes harder and warmer and it will not spontaneously return to its softer, undeformed state. If we break a china cup into pieces it cannot be returned to its former pristine state by carefully putting all the pieces back together again. If we discharge a capacitor through a resistor the charge decreases and eventually reaches zero. Heat flows from regions of high temperatures to regions of low temperatures but not back again. It is the irreversibility of these processes that defines our sense of the direction of time. Although the fundamental equations of motion of atoms and molecules are symmetric with respect to the reversal of time, the Universe is not.

In thermodynamics the degree of irreversibility of a process is characterised and quantified by something called *entropy*. In all irreversible processes the entropy of the system and its surroundings increases. For a process to be reversible the total change of entropy has to be zero. If there is a negative change in entropy somewhere it is always compensated by a positive change of entropy at least as large somewhere else. The second law of thermodynamics may be stated in equivalent ways as follows:

- *No process is possible in which the only result is the complete conversion of heat into work.*

- *No process is possible in which the only result is the transfer of heat from a colder to a hotter body.*

In a *spontaneous process* the system changes in a natural way from a non-equilibrium state towards an equilibrium state. It is only when the system is in equilibrium that the entropy is constant. If the system is isolated its entropy is then a maximum. If this were not true a spontaneous change in the isolated system would increase its entropy, and therefore the system could not have been at equilibrium. Thermodynamics tells us only about the direction of change, and the ultimate destination. The time it takes for these changes to occur is not predicted by equilibrium thermodynamics.

---

[5] If there were an *infinite* number of possible trajectories for the 10^{30} molecules then the probability of the man emerging from the water in a dry state and returning to his original position would be $1/\infty = 0$. The reason it is not zero, albeit extremely small, is that there is a lower limit on the difference between two trajectories for them to be classified as distinct. The limit is set by Heisenberg's uncertainty relation of quantum theory. It states that the uncertainty in a measurement of a position coordinate of a particle multiplied by the uncertainty in a measurement of the corresponding momentum coordinate is at least equal to the Planck constant, $h = 6.626 \times 10^{-34}$ J s. This discretises the 6×10^{30} dimensional space (each molecule has 3 position coordinates and 3 momentum coordinates) of the molecular trajectories into 10^{30} cells, each of volume h^3. If the positions and momenta of a given particle fall within the same h^3 cell they must be treated as the same. Although h is small it is not zero. It follows that the number of possible trajectories of each molecule is a very large number but it is not infinite. (This is the reason why the Planck constant appears in the statistical mechanics of particles obeying classical physics, even though it is normally a signature of quantum physics.)

A *reversible process* takes the system through a continuous sequence of equilibrium states. This is an unattainable limiting process that can be approached in reality only by maintaining the system extremely close to equilibrium. Such a process can be exactly reversed if it is driven very slightly in the reverse direction. The process has to be carried out extremely slowly to allow the whole system to re-equilibrate after each extremely small change. The change of entropy associated with a reversible process is zero because the system passes through a sequence of only equilibrium states.

The change of entropy of an object which undergoes a change of equilibrium state, such as a phase change or a change of temperature, is independent of whether the change occurs reversibly or irreversibly. That is because the entropy of the object is uniquely defined by the variables that define its equilibrium state, such as temperature, pressure and chemical potentials. In a change of equilibrium state the change of entropy depends only on the initial and final states, and not on how it gets from the initial to the final equilibrium states. This property of entropy elevates it to a *function of state*. Any property of the system which is uniquely defined by its equilibrium state is called a function of state. The internal energy is another example of a function of state.

In a reversible change of state the change of entropy of the surroundings is the negative of the change of state of the object. The change of the total entropy is then zero. In an irreversible change of state of a system the sum of the changes of entropy in the object and its surroundings is always greater than zero, by an amount that increases with the degree of irreversibility. But in all cases the change of entropy of the object undergoing a change of state is the same irrespective of the degree of irreversibility of the change.

If a small quantity of heat δq is transferred to an object reversibly the increase of the entropy of the object is *defined* by $\delta q/T$, where T is the temperature of the object. In this definition δq has to be small in general because otherwise the temperature of the object will change after δq has been added. But if the heat transferred is latent heat associated with a phase change then δq can be the entire latent heat since the temperature T is constant during the phase change. If the object loses heat δq is negative and the entropy of the object decreases.

Since the internal energy of an object is a function of state it changes by the same amount in a change of state, irrespective of whether the change occurs reversibly or irreversibly. In general a change of state involves both the addition or removal of heat and work being done on or by the object[6]. Only the sum of these two contributions to the change of internal energy is independent of whether the change occurs reversibly or irreversibly. This means that in general the heat gain or loss and the work done on or by the object are not functions of state.

Suppose we have an isolated system in which there are local variations in temperature. If a small amount of heat δq leaves a local region with temperature T_1 the entropy of the region changes by $-\delta q/T_1$. If the quantity of heat δq is transferred to

---

[6]A Joule expansion is an interesting exception to this statement: an ideal gas occupies half a container and is separated from a vacuum in the other half by a removable partition. The entire container is thermally insulated. The partition is removed and the gas quickly occupies the whole container. No heat enters or leaves the gas and no work is done on or by the gas. Its internal energy and its temperature remain the same. However, as we shall see in the next section its entropy increases because it occupies twice the volume.

a region where the temperature is T_2 its entropy changes by $+\delta q/T_2$. The change of the total entropy of the system is $\delta q/T_2 - \delta q/T_1$. The second law says that for this process to occur spontaneously the total change of entropy must be positive. That is true provided $T_1 > T_2$. In other words, heat flows spontaneously only from regions of higher temperature to regions of lower temperature, which accords with our experience. The state of maximum entropy is when the temperature is constant throughout the system. We recognise this as a condition for thermal equilibrium in an isolated system.

The second law recognises that there is something unique about heat. It is quite easy to convert potential energy into electrical energy, and to convert electrical energy back into potential energy, albeit with some losses due to friction. This is what is done frequently at Dinorwig power station[7] in Snowdonia, North Wales. Conversions between other forms of energy are also feasible. But not when the final product is heat. For example, when an aircraft lands most of its kinetic energy ends up as heat in its brakes. That heat cannot be used to send the aircraft back into the air, despite the first law. The second law and our experience tell us that when energy is converted into heat it is impossible to convert all that heat back into work. This is often described as the 'degradation' of energy. As we shall see in the next section statistical mechanics explains this property of heat in terms of the *dispersal* of energy among microstates of the system. As energy becomes more dispersed it becomes less capable of doing work, such as propelling an aircraft into the air.

1.5.2 Entropy in terms of microstates

We have seen in section 1.3 that the thermodynamic state of an isolated system is expressed in terms of state variables such as pressure, volume and temperature. These are macroscopic variables and they define the *macrostate* of the system. As I discuss in Chapter 3, if we were able to look inside one of these macrostates at the atomic scale we would see atoms in constant motion. If there are N atoms in the isolated system there are $3N$ variables associated with their positions and another $3N$ variables associated with their instantaneous velocities[8]. These $6N$ variables constitute a *microstate* of the system. To each macrostate of the system there is a very large number of possible microstates[9]. But some macrostates may have many more microstates than others. For example when a crystal melts the liquid state has many more microstates than the crystal because the atoms are no longer confined to their average positions within the crystal.

The concept of entropy was introduced at a time when not everyone believed matter was made of atoms. Its derivation in classical thermodynamics involves 'Carnot cycles' and 'heat engines' and it is somewhat abstract. It was Ludwig Boltzmann who provided a more physical understanding of entropy. Boltzmann showed that entropy increases with the number W of *microstates* of an isolated system, given that the internal energy, numbers of particles of each species and volume and shape of the

---

[7] https://www.electricmountain.co.uk/Dinorwig-Power-Station

[8] The factors of 3 arise from the 3 spatial dimensions of the system.

[9] Although the number of microstates is very large it is not infinite for the quantum mechanical reason explained in footnote 5.

system are all constant. If it is assumed that each microstate is equally probable Boltzmann showed that the entropy S is proportional to the logarithm of the number W of microstates of the system:

$$S = k_B \log_e W \tag{1.1}$$

where the constant of proportionality k_B is the *Boltzmann constant* equal to 1.381×10^{-23} Joules per kelvin ($\mathrm{J\,K^{-1}}$). The Boltzmann constant is the gas constant[10], $R = 8.314\,\mathrm{J\,K^{-1}}$, divided by Avogadro's number, $N_A = 6.022 \times 10^{23}$. If there is an increase in entropy associated with a change of macrostate of an isolated system, the new macrostate has a larger number of microstates in which it can exist. This is what is meant by saying that the internal energy is dispersed over a larger number of microstates when the entropy increases. It does not mean that at any given instant in time the internal energy is distributed among more coexisting microstates. It means there are more microstates available from which one is selected for the entire macrostate at any given instant in time.

It is not difficult to see why the entropy has to depend on the *logarithm* of the number of microstates. Consider two isolated systems A and B at the same temperature. Let the number of microstates of system A be W_A and of system B be W_B. The entropies of the separate systems are $S_A = k_B \log_e W_A$ and $S_B = k_B \log_e W_B$. Suppose the two systems are brought into thermal contact to make a single combined system, while maintaining a barrier between them to prevent their contents from mixing. Since the two systems A and B are at the same temperature no heat flows between them. With no heat flow between them, and no intermixing of their contents, the entropy of the combined system is the same as the total entropy of the separate systems A and B. The total entropy of the separate systems is $S_A + S_B = k_B \log_e W_A + k_B \log_e W_B$. The number of microstates of the combined system is $W_A W_B$, and therefore its entropy is $k_B \log_e(W_A W_B)$. These expressions are indeed equal to each other because $\log_e(W_A W_B) = \log_e W_A + \log_e W_B$. Only the logarithm has this property.

In an ideal gas at constant temperature the number of microstates available is proportional to the volume of the gas because each gas particle is free to roam throughout the volume. Therefore the entropy of an ideal gas at constant temperature is proportional to the logarithm of the volume it occupies. This is why the entropy of the gas in a Joule expansion increases (see footnote 6). In contrast, atoms in a solid are confined by their neighbours to move in much smaller volumes than the total volume of the solid. They vibrate about their equilibrium positions, and the microstates available to them at constant temperature increase with their amplitudes of vibration. Therefore, atoms in less confined, more open spaces in a given solid contribute more to the total entropy of the solid than atoms in more confined environments. These ideas carry over to exotic materials such as colloids where particles that are small but much larger than atoms are held in suspension in a liquid. It is found that inert spheres suspended in

---

[10]the gas constant is R in the equation of state of an ideal gas $PV = nRT$, where P is the gas pressure measured in pascals, V its volume in cubic metres, T its temperature in kelvin and n is the amount of the gas in moles.

a liquid may sometimes crystallise in structures that are not densely packed. These open structures are stabilised not by potential energy but by their higher entropy[11].

1.5.3   Configurational entropy

The entropy of an isolated system may also increase as the result of mixing distinguishable atoms together. This is called *configurational entropy*. As a simple example, consider a hypothetical square two-dimensional crystal with a square lattice, comprising 10×10 lattice sites. For simplicity we shall ignore the atomic velocities in this example. Each of the 100 lattice sites may be occupied by either a black atom or a white atom. The colour denotes either a distinct isotope of the same element, such as carbon-12 and carbon-14, or atoms of two elements. Let there be 50 black atoms and 50 white atoms arranged on the 100 sites. There is a 'maximally separated' state as shown in Fig. 1.1a. There is also a 'maximally intermixed' state where the four neighbours of each black atom are white and *vice versa*, as shown in Fig. 1.1b. Fig. 1.1c illustrates a configuration where the 100 atomic sites are occupied randomly by 50 black and 50 white atoms. It is an amazing fact that there are around 10^{29} similar random configurations of 50 black and 50 white atoms on these 100 lattice sites[12]. Each of the configurations shown in Fig. 1.1 is a microstate of the crystal of 100 atoms.

If the black and white atoms are isotopes of the same atom the potential energies of all these configurations are equal, because the energies of the bonds between atoms are independent of the number of neutrons in each atomic nucleus. At all temperatures up to the melting point the entropy is maximised when the crystal is disordered because all 10^{29} configurations are equally accessible, and the overwhelming majority of them are disordered. In a real crystal with many more than 100 atomic sites the number of possible random atomic configurations increases rapidly. This is why it was so difficult to separate the isotopes of uranium in the Manhattan Project to develop the atomic bomb.

Suppose bonds between atoms of the same colour are of lower potential energy than those between atoms of different colours. The potential energy of the system is minimised when the number of bonds between black and white atoms is minimised. This corresponds to the configuration of Fig. 1.1a. There are a further three configurations equivalent to Fig. 1.1a obtained by rotating the whole crystal about the normal to the page by $90°, 180°$ and $270°$. Any deviation from the configuration shown in Fig. 1.1a requires energy, but if there is no heat in the isolated system no such deviations are permitted by the first law. The configurational entropy is then $k_B \log_e 4$ because there are four equivalent configurations.

Imagine we heat the system depicted in Fig. 1.1a rapidly to a low temperature and then isolate it. The new equilibrium state of the isolated system is determined by maximising the entropy at the slightly higher internal energy. An exchange of neighbouring black and white atoms at the interface raises the potential energy of the system because it increases the number of bonds between atoms of different colours

---

[11] Mao, X, Chen, Q and Granick, S, Nature Mater **12**, 217 (2013).

[12] The total number of configurations is the binomial coefficient $^{100}C_{50} = 100!/(50!)^2$. The number of random configurations is 6 less than $^{100}C_{50}$ because there are 4 configurations like Fig. 1.1a and 2 configurations like Fig. 1.1b.

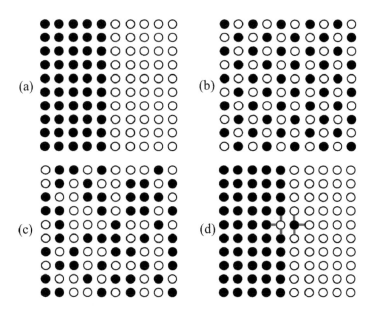

Fig. 1.1 Four atomic configurations of a 10×10 square crystal lattice, with 50 'black' atoms and 50 'white' atoms occupying the 100 lattice sites. (a) the most separated state. (b) the most intermixed state. (c) one of the $\approx 10^{29}$ configurations with sites occupied by black and white atoms at random. (d) the atomic configuration shown in (a) with one exchange between black and white atoms at the interface. The exchange introduces 6 new bonds, shown in red, between black and white atoms.

by six, as shown in Fig. 1.1d. Provided there is sufficient heat in the crystal to supply this increase of potential energy a local thermal fluctuation can enable the exchange to occur, at least in principle. The additional potential energy of the new bonds is accompanied by a reduction in the kinetic energy of the system, maintaining a constant value of the internal energy. The pair of exchanged atoms at the interface may proceed to exchange with other atoms with no further increase of potential energy. Thus, once an exchange has occurred at the interface the exchanged white atom can occupy any of the 50 black atom sites of Fig. 1.1d. Similarly, the exchanged black atom can occupy any of the 50 white atom sites. Thus one exchange at the interface leads to $50 \times 50 = 2,500$ new configurations that share the same potential energy. The configurational entropy has increased as a result of this one exchange at the interface.

Further exchanges can take place if there is sufficient heat in the system. At equilibrium the atomic structure of the system is not constant in time but is running through all the accessible states at the given internal energy. As the atomic structure changes the total potential energy of the bonds changes too. The temperature of the system has to change to maintain constant the internal energy. The average temperature will settle down to a value at equilibrium that is less than the temperature of the system before it equilibrated. Further injection of heat into the system will eventually enable all 10^{29} configurations to become accessible. Although the state shown in Fig. 1.1a, or one of the three equivalent configurations, continues to have the lowest potential energy it is extremely unlikely to arise when there are so many more accessible disordered configurations like Fig. 1.1c.

The 'most intermixed' configuration shown in Fig. 1.1b arises at low temperatures when bonds between atoms of different colours have a lower potential energy than those between atoms of the same colour[13]. As the crystal is heated an increasing number of atomic exchanges may take place until eventually all 10^{29} random configurations become accessible, assuming it does not melt first. Thus the equilibrium configuration of the crystal at high temperatures is an average taken over all 10^{29} configurations, almost all of which are random. In this state if the atoms really were coloured black and white they would all appear grey in a time-lapse photograph – half way between black and white.

1.5.4   Summary of section 1.5

The irreversibility of spontaneous, natural processes imposes a direction on time. Such processes are always associated with an increase of entropy in the system and its surroundings: this is the second law of thermodynamics. The greater the increase of entropy the greater the degree of irreversibility. Only reversible processes create no entropy. In an isolated system entropy is directly related to the number of microstates the system can access with the internal energy the system has. The irreversibility of converting work into heat is caused by the dispersal of the energy among a very large number of microstates of the system. This dispersal is sometimes called the 'degradation of energy': energy is conserved when it is converted into heat but it is less capable of doing work.

1.6   Closed systems and heat reservoirs

So far we have considered the entropy of only isolated systems. In a closed system the object we are studying can exchange heat with the environment, but it cannot exchange matter. To treat a closed system we imagine the object of study is in thermal contact with a large heat reservoir at a constant temperature T_r. The role of the heat reservoir is to accept or donate heat to keep the object at a constant temperature of T_r. In the process it is assumed there are no irreversible changes of entropy within the reservoir such as mixing or chemical reactions. The reservoir is in a state of internal equilibrium. Under these conditions if the heat reservoir donates a small amount of heat δq_{rev} to the object it undergoes a reversible change of state and the entropy of the reservoir changes by $\delta S_{res} = -\delta q_{rev}/T_r$. If the object accepts the heat δq_{rev} reversibly its entropy changes by $+\delta q_{rev}/T_r$. Thus, in a reversible transfer of heat the total change of entropy is zero.

In a reversible transfer of heat δq_{rev} from the reservoir to the object the internal energy of the object increases by $\delta U = \delta q_{rev}$. That is because no work is done by or on the object when the change of state involves a transfer of heat only. But suppose

---

[13]Note there are black atoms at the top left and bottom right corners and white atoms at the top right and bottom left corners. There is an equivalent configuration, obtained by rotating the crystal by $\pm 90°$ about the normal to the page, where the positions of all the black and all the white atoms are interchanged. Therefore, the entropy of this configuration is $k_B \log_e 2$.

the same change of internal energy $\delta U = \delta q_{rev}$ of the object is brought about by an *irreversible* change of its state in which work δw is done on it in addition to a transfer of heat δq_{irrev}. Therefore, $\delta U = \delta q_{rev} = \delta q_{irrev} + \delta w$. The change of entropy of the reservoir is $\delta S_{res} = -\delta q_{irrev}/T_r$ because it has lost heat δq_{irrev} and all changes of heat content of the reservoir occur reversibly. Since the object has undergone the same change of state its change of entropy remains $\delta S_{obj} = \delta q_{rev}/T_r$ because entropy is a function of state. Since the change of state is now irreversible it must be true that $\delta S_{res} + \delta S_{obj} > 0$, and therefore $\delta q_{rev} > \delta q_{irrev}$. The difference $\delta q_{rev} - \delta q_{irrev}$ is provided by the work δw done on the object which is degraded to heat in the object.

On the other hand, if the change of state involves a transfer of heat from the object to the reservoir, more heat is transferred if the change of state occurs irreversibly than if it occurs reversibly. Let δq_{rev} and δq_{irrev} be the amounts of heat transferred from the object to the reservoir in reversible and irreversible changes of state respectively. The change in the entropy of the object is $\delta S_{obj} = -\delta q_{rev}/T_r$ regardless of whether the change of state occurs reversibly or irreversibly. The change of entropy of the reservoir is $\delta S_{res} = +\delta q_{irrev}/T_r$. Therefore, the total change of entropy is $(\delta q_{irrev} - \delta q_{rev})/T_r$ which must be positive. The difference $\delta q_{irrev} - \delta q_{rev}$ is the work done by the object that is degraded to heat in the reservoir.

In both cases work is degraded to heat when the change of state occurs irreversibly.

1.7   The Helmholtz free energy

We have seen that in an isolated system equilibrium is achieved when the entropy of the system is maximised. As an isolated system evolves towards equilibrium its temperature changes. It is more useful to have a criterion for equilibrium in a closed system where the temperature is kept constant by placing the object in contact with a thermal reservoir. However, no particles leave or enter the object. Nevertheless, there can still be movement of particles and chemical reactions within the object resulting in new phases. The equilibrium criterion is provided by the Helmholtz free energy, which is a function of internal energy, entropy and temperature and therefore it is a function of state.

In case you are wondering whether 'free' energy is the answer to the problems of the World's energy supply there is no such thing as energy without an economic cost. The use of the word 'free' in this context is quite different. As we shall see it means the maximum energy available to do work in a system held at a constant temperature. Better descriptors would be 'available work' or 'useful energy', but we are stuck with 'free energy' because it has become common usage.

The Helmholtz free energy A is defined as the internal energy U minus the temperature T times the entropy S, thus $A = U - TS$. Consider an object immersed in a large heat reservoir at a temperature T_r. Let the object undergo a change from a state labelled '1' to a state labelled '2'. The Helmholtz free energy of the object in state 1 is $A_1 = U_1 - T_r S_1$ and in state 2 it is $A_2 = U_2 - T_r S_2$. Let the amount of heat entering the object in going from state 1 to 2 be q. Let the work done by the object in going from state 1 to 2 be w. Then $U_2 - U_1 = q - w$. Since the internal energy is a function of state, $U_2 - U_1$ is independent of the degree of irreversibility of the change of state. But q and w do depend on the degree of irreversibility in such a way that

$q - w$ is the same as when the change of state occurs reversibly. If the change of state occurs reversibly then we saw in the previous section that q is a maximum and it is equal to $T_r(S_2 - S_1)$. The difference in Helmholtz free energies is then $A_2 - A_1 = -w$, which may be written as $w = A_1 - A_2$. But if the change of state occurs irreversibly we saw in the previous section that the amount of heat q is less than $T_r(S_2 - S_1)$. In that case $A_2 - A_1 < -w$, which may be written as $w < A_1 - A_2$. Thus the change in the Helmholtz free energy between the two states is the maximum amount of work available from the change. The maximum is attained when the change of state occurs reversibly.

Returning to the change in the Helmholtz free energy $A_2 - A_1 < -w$, we see that $A_2 - A_1 < 0$ in a spontaneous change of state if there is no work w done. It follows that the Helmholtz free energy evolves towards a minimum value in a closed system where the temperature is held constant and it is held in a rigid container so that no work is done by or on the object. Minimisation of the Helmholtz free energy is the criterion for equilibrium of the system under these conditions.

1.8   The Gibbs free energy

At last we arrive at one of the most useful thermodynamic constructs for materials science (and chemistry). In the previous section we saw that minimisation of the Helmholtz free energy is the criterion for equilibrium in an object held at constant temperature provided the surfaces of the object are constrained so that no work can be done by or on the object. If the object has a tendency to expand as it approaches the equilibrium state the expansion is suppressed by the constraints imposed on it and the pressure in the object rises. In materials science we are normally concerned with materials at the constant pressure provided by the Earth's atmosphere of $101\,\text{kPa}$ at sea level. In most solids this pressure has a negligible influence on thermodynamic stability. But in soft condensed matter and when gases are involved it is important to have a criterion for equilibrium in systems held at constant pressure as well as constant temperature. In some deformation processes in metals and alloys sufficiently high pressures may be created which stabilise phases that would not normally be stable at atmospheric pressure. In earth sciences we may want to know which phases of a mineral are stable over a range of pressures and temperatures, such as those that may exist in the lithosphere. Minimisation of the Gibbs free energy is the criterion for equilibrium in a system maintained at constant temperature and constant pressure.

The Gibbs free energy, G, is defined as the Helmholtz free energy $U - TS$ plus the pressure P times the volume V, thus $G = U - TS + PV$. Since T, P and V are state variables and U and S are functions of state, the Gibbs free energy is a function of state.

Consider the change in the Gibbs free energy of an object from a state labelled '1' to a state labelled '2' at constant pressure P and constant temperature T. Then $G_2 = U_2 - TS_2 + PV_2$ and $G_1 = U_1 - TS_1 + PV_1$. As in the previous section we can write $U_2 - U_1 = q - w$, which is independent of whether the change of state occurs reversibly or irreversibly. Since $q \leq T(S_2 - S_1)$ we have $G_2 - G_1 \leq -w + P(V_2 - V_1)$. The term $P(V_2 - V_1)$ is the mechanical work done by the object against the applied pressure P in changing the volume of the object from V_1 to V_2, and it is included in w. As we shall see in the next section there are chemical forms of work that can take place within an object that contribute to w but do not necessarily contribute to a change of volume. To take these other forms of work into account we write $w = w_c + P(V_2 - V_1)$, where w_c represents the chemical work terms. Then $w_c \leq G_1 - G_2$. In a process in which the temperature and pressure are maintained constant throughout the system, the work done by chemical processes in the object is less than or equal to the decrease of the Gibbs free energy. When there are no chemical changes within the object $G_2 - G_1 \leq 0$. *The criterion for equilibrium of a system held at constant temperature and constant pressure is that the Gibbs free energy is a minimum.* These are the conditions under which experiments on materials are usually conducted, which is why the Gibbs free energy is central to phase equilibrium in materials.

1.9 Chemical potentials

In an isolated system the internal energy is constant, no mechanical work is done by or on the system, and no heat enters or leaves the system. But it would be wrong to infer there can be no change in the entropy of an isolated system. For example, there could be an increase in configurational entropy as a result of mixing of the components, or changes of phase could occur generating heat. Another example of a change of entropy in an isolated system is a Joule expansion, described in footnote 6.

In a homogeneous material the internal energy, Helmholtz free energy and Gibbs free energy are extensive quantities. Therefore, they must depend on the numbers of particles of each substance present in the system. At constant temperature and pressure the change, δG, in the Gibbs free energy when the number n_i of particles of component i changes by a small amount δn_i, while the numbers of all other components are held constant, is $\mu_i \delta n_i$. This is the most practical definition of the chemical potential μ_i of component i. The chemical potential is an intensive property of the system like temperature and pressure. Differences in temperature drive the transport of heat. Differences in pressure drive changes of volume. Differences in chemical potential drive transport of particles. At equilibrium the temperature and pressure and the chemical potential of each component are all constant throughout the system[14].

Consider a single component substance in equilibrium with its vapour, such as water. At a given temperature and pressure the vapour pressure is determined by equality of the chemical potential of water molecules in the vapour phase and in the liquid phase. If water molecules are withdrawn from the vapour the chemical potential in the vapour phase becomes less than in the liquid phase. Water evaporates from the liquid to restore equilibrium. In a multi-component material the chemical potential of each component may be manipulated by varying the pressure of the component in a gas with which the material is in equilibrium. Variations over many orders of magnitude of the pressure of a component in the gas phase may be achieved by placing the material of interest in a sealed vessel with another material where the chemical potential of the component is very different. For example, by putting a metal oxide in a sealed vessel with another material with a more negative chemical potential for

---

[14]When fields are present in the material, such as stress fields or magnetic fields in a magnetic material, they have to be taken into account in the condition for equilibrium.

oxygen (which means a stronger affinity for oxygen) the metal oxide may be reduced to the metal.

Consider a material containing k components in thermodynamic equilibrium with its vapour. Let us assume the components are atomic not molecular. Let the number of atoms of component j in the material be n_j. The atomic concentration of component j in the material is then $c_j = n_j/N$ where N is the total number of atoms in the material. If the material grows reversibly at constant temperature and pressure by accreting an additional total number of atoms ΔN from the vapour, the increase in the number of atoms of component j is $c_j \Delta N$. To maintain equilibrium in the system the chemical potential of each component remains constant during the accretion process. The change in the Gibbs free energy of the material is then $\Delta G = (\mu_1 c_1 + \mu_2 c_2 + \cdots + \mu_k c_k)\Delta N$. By continuing the accretion process we deduce the free energy per atom g in the material is $g = \mu_1 c_1 + \mu_2 c_2 + \cdots + \mu_k c_k$. Here we see the explicit dependence of the Gibbs free energy of the material on the chemical composition.

Let us repeat the argument but for the internal energy this time. Consider the reversible growth of a material containing k components by accreting a small number δn_j of atoms of each component j in such a way that the composition of the material is constant. The temperature T, pressure P and chemical potentials $\mu_1, \mu_2, \ldots, \mu_k$ are constant throughout the system. From the definition $\delta S = \delta q/T$ of the change of entropy δS associated with a change of heat δq, the change in the heat of the material is $T\delta S$. The mechanical work done by the material is $P\delta V$, where δV is the small change of volume of the material. The change in the internal energy of the material associated with the increase in the numbers of atoms of each component is $(\mu_1 \delta n_1 + \mu_2 \delta n_2 + \cdots + \mu_k \delta n_k)$. Therefore, the change in the internal energy of the material is:

$$\delta U = T\delta S - P\delta V + \mu_1 \delta n_1 + \mu_2 \delta n_2 + \cdots + \mu_k \delta n_k. \tag{1.2}$$

This equation is called the combined first and second laws of thermodynamics. It follows that the change in the internal energy at constant entropy and volume when the number of atoms of component j changes by δn_j is $\mu_j \delta n_j$. This is an alternative, but not very practical, definition of the chemical potential. The more important point is that the terms after $T\delta S$ represent work. The term $(\mu_1 \delta n_1 + \mu_2 \delta n_2 + \cdots + \mu_k \delta n_k)$ is 'chemical work' and it corresponds to the term w_c in the previous section.

1.10   The Gibbs-Duhem equation

By continuing the accretion process that led to eqn 1.2 we find the internal energy U is equal to $TS - PV + \mu_1 n_1 + \mu_2 n_2 + \cdots + \mu_k n_k$. This equation follows because the entropy, volume and numbers of atoms of each species are all extensive variables, whereas the temperature, pressure and chemical potentials are constant. Consider the change of the internal energy if the temperature, pressure and the chemical potentials are changed by small amounts in addition to small changes in entropy, volume and numbers of atoms of each component. We obtain[15]:

---

[15] Consider the change $\delta(xy)$ of xy if x changes to $x + \delta x$ and y changes to $y + \delta y$, where $\delta x/x \ll 1$ and $\delta y/y \ll 1$. Then $\delta(xy) = (x + \delta x)(y + \delta y) - xy$. Multiplying the terms in brackets we find $\delta(xy)$

$$\delta U = (T\delta S - P\delta V + \mu_1 \delta n_1 + \mu_2 \delta n_2 + \cdots + \mu_k \delta n_k)$$
$$+ [S\delta T - V\delta P + n_1 \delta\mu_1 + n_2 \delta\mu_2 + \cdots + n_k \delta\mu_k].$$

Comparing this with eqn 1.2 we see the term in square brackets must be zero, that is:

$$S\delta T - V\delta P + n_1 \delta\mu_1 + n_2 \delta\mu_2 + \cdots + n_k \delta\mu_k = 0. \qquad (1.3)$$

This relation is called the Gibbs-Duhem equation. It says that at equilibrium variations in the temperature, pressure and chemical potentials are not independent. There is a Gibbs-Duhem equation for each phase present in the system at equilibrium.

1.11   The Gibbs phase rule

The Gibbs phase rule is central to the construction of phase diagrams, which I describe in the next chapter. Consider two phases α and β consisting of the same single component κ. For example, they could be liquid water and ice. Let the phase α be at the temperature T_α and pressure P_α. The chemical potential of the component κ in the phase α is determined: $\mu_\kappa^{(\alpha)} = \mu_\kappa^{(\alpha)}(T_\alpha, P_\alpha)$. Let the phase β be at the temperature T_β and pressure P_β. The chemical potential of the component κ in the phase β is determined: $\mu_\kappa^{(\beta)} = \mu_\kappa^{(\beta)}(T_\beta, P_\beta)$. If the two phases are to coexist in equilibrium we must have $T_\alpha = T_\beta$, $P_\alpha = P_\beta$ and $\mu_\kappa^{(\alpha)} = \mu_\kappa^{(\beta)}$. Therefore, there are three equations relating the four variables $T_\alpha, P_\alpha, T_\beta$ and P_β. There is 1 degree of freedom (because $4 - 3 = 1$). This means there is only one state variable that may be varied at will. But once it is fixed the thermodynamic state of the system comprising the two phases α and β in equilibrium is completely determined. For example, at normal atmospheric pressure of 101 kPa the temperature at which ice and water are in equilibrium is 0°C. But at a pressure of 209.9 MPa the melting point of ice changes to −21.9°C.

If there are three phases of a single component in equilibrium the temperature and pressure of each phase amount to six variables. At equilibrium there are two equations relating the three temperatures, two equations relating the three pressures and two equations relating the three chemical potentials. Therefore, there are $6 - 6 = 0$ degrees of freedom: there is a unique temperature and pressure at which three phases of a single component system may be in equilibrium. It is known as the triple point. In pure water the triple point is at the temperature 0.01°C and pressure 611.2 Pa.

If there are P phases in equilibrium in a multi-component system of C components the number of degrees of freedom F is given by:

$$F = C + 2 - P. \qquad (1.4)$$

This is the Gibbs phase rule[16] . The number of thermodynamic state variables is $C+2$ because there are C chemical potentials in addition to temperature and pressure.

---

is $(xy + x\delta y + y\delta x + \delta x\delta y) - xy = x\delta y + y\delta x + \delta x\delta y$. Therefore, the fractional change $\delta(xy)/(xy)$ is $\delta x/x + \delta y/y + (\delta x/x) \times (\delta y/y)$. Since the first two terms are much less than 1 the third term is very much smaller. Therefore to a very good approximation $\delta(xy)/(xy)$ is $\delta x/x + \delta y/y$. Multiplying both sides by xy we obtain $\delta(xy) = x\delta y + y\delta x$. As δx and δy tend to zero this equation becomes exact. It has been used to derive δU. For example, the change is TS is $\delta(TS) = T\delta S + S\delta T$.

[16]David Gaskell, who taught me thermodynamics when I was a student at the University of Pennsylvania, gave the following mnemonic to remember the phase rule:
a **P**olice **F**orce = a **C**onstable + **2**

However, for each of the P phases there is a Gibbs-Duhem relation between the $C+2$ variables. Hence the rule.

1.12   Closing remarks

This chapter has covered a great deal of ground from the laws of thermodynamics to the Gibbs phase rule. The phase rule underpins the construction of temperature-composition phase diagrams which we move onto in the next chapter. But we have barely touched the surface of the role of thermodynamics in materials science. Another of its applications, which involves the calculus of several variables, is to establish often surprising relationships between properties of materials that would be difficult to deduce otherwise. These are particularly useful when a property that is extremely difficult to measure experimentally is expressed in terms of other quantities that are measurable.

Thermodynamics is not concerned with the mechanisms of any changes of phase or how long they take to reach the equilibrium state. To determine the rates at which materials undergo phase changes and approach equilibrium we have to consider 'kinetics'. In practice, since materials are not usually at equilibrium, kinetics is as significant as thermodynamics in determining the phases present. It involves an examination of the atomic mechanisms by which change occurs in materials. We shall discuss some of those mechanisms in Chapter 4.

Further reading

Adkins, C J, *Equilibrium thermodynamics*, 3rd edition, Cambridge University Press (1987).

Cottrell, A H, *An Introduction to metallurgy*, 2nd edition, The Institute of Materials (1995).

Denbigh, K, *The principles of chemical equilibrium*, Cambridge University Press (1964).

Gaskell, D R and Laughlin, D E, *Introduction to the Thermodynamics of Materials*, Taylor and Francis (2018).

2

Phase diagrams

What is the use of a book without pictures?

Lewis Carroll in *Alice's Adventures in Wonderland*

2.1 Introduction

One of the most useful applications of thermodynamics in materials science is the construction of phase diagrams. In this chapter we will see how to construct temperature - composition phase diagrams for two-component systems, known as binary systems.

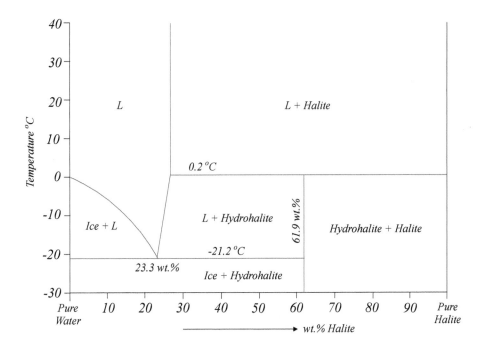

Fig. 2.1 The phase diagram between water and halite (common salt). *L* denotes liquid.

Let's dive straight in with an example of a temperature-composition phase diagram. Figure 2.1 is the phase diagram for water and halite. Halite is common salt – NaCl. The phase diagram is a map of the stable phases, at atmospheric pressure, that may be formed at different temperatures in mixtures of water and halite. Pure water is on the far left and pure halite is on the far right. The concentration is expressed as a percentage of halite by weight. This means that in a 100 gram sample of 30 wt.% halite there are 30 grams of halite and 70 grams of water[1].

On the far left we see the familiar freezing point of pure water at 0°C. If we reduce the temperature below 0°C all of the pure liquid water transforms to pure ice. There is a unique equilibrium freezing point of pure water at atmospheric pressure and it is 0°C. But suppose we cool a liquid solution of 10 wt.% halite dissolved in water. This solution is called brine. The blue line separating the liquid solution from the mixture of ice and liquid bends downwards to lower temperatures. At 10 wt.% halite it is crossed at about −6°C. At that point we have a mixture of very small ice crystals in brine. We put salt on the roads in winter because the freezing point of water is reduced by the salt. If we continue to reduce the temperature something very interesting happens. The ice crystals continue to grow as essentially pure ice because the ice crystal lattice accepts only an extremely small concentration of sodium and chlorine ions – so small it does not show in the phase diagram. Consequently the concentration of halite in the remaining liquid must increase, following the blue line separating the L and $Ice + L$ fields. This process continues as we reduce the temperature further until we reach the temperature of −21.2°C when the remaining liquid finally crystallises to a mixture of ice and hydrohalite. Hydrohalite is a solid hydrated form of halite with the formula $NaCl \cdot 2H_2O$. Thus there is a range of temperatures between about −6°C and −21.2°C over which the 10 wt.% halite brine solution gradually freezes to a mixture of ice and hydrohalite.

If we cool a liquid solution of 23.3 wt.% halite in water it remains liquid to a temperature of −21.2°C. At that unique temperature the liquid, ice and hydrohalite coexist in equilibrium. If we reduce the temperature slightly *all* the liquid is transformed to ice and hydrohalite. At the unique concentration of 23.3 wt.% halite there is a unique freezing point. This is an example of a eutectic phase change, which I will describe in more detail in Section 2.5.1. For now I note it is a remarkable phase change because the 23.3 wt.% brine has to separate into solid ice, where there is virtually no halite, and solid hydrohalite, which has a relatively high concentration of halite. For that to happen there has to be considerable movement of the sodium and chlorine ions in the brine.

On the far right side we have pure halite, which is a solid in the range of temperatures shown in the diagram. As we add water to pure halite at temperatures below 0.2°C we form a solid mixture of halite and hydrohalite. As we continue to increase

---

[1]To convert this to a percentage concentration of molecules of halite c_{halite} we need the molar masses of halite and water, $M_{halite} = 58.44$ grams per mole and $M_{water} = 18.00$ grams per mole. The percentage concentration of molecules of halite in a sample of 30 wt.% halite is then:

$$c_{halite} = 100 \times \frac{\frac{30}{M_{halite}}}{\frac{30}{M_{halite}} + \frac{70}{M_{water}}} = 11.7\%.$$

the concentration of water at the same temperature we eventually reach a state at 61.9 wt.% halite where there is only pure solid hydrohalite. What happens next if we continue to decrease the halite concentration depends on the temperature. At temperatures below $-21.2°C$ we form a mixture of ice and hydrohalite crystals. Above $-21.2°C$, but still below $0.2°C$, we form hydrohalite crystals in a saturated brine solution. Above $0.2°C$ the hydrohalite decomposes into water and halite, so that we have halite crystals in a saturated brine solution.

Phase diagrams such as Fig. 2.1 are extremely useful. Not only do they tell us which phases are present at a given temperature and composition, but the boundaries (the blue lines in Fig. 2.1) show where phase changes occur. Thus, there is a great deal of information encapsulated in a readily digestible form in a phase diagram. In the rest of this chapter we will consider how simple phase diagrams are constructed using thermodynamic principles described in the previous chapter.

2.2   Free energy - composition curves

As in chapter 1, we label the two atomic components of the binary system B and W, short for black and white. Let the fraction of atomic sites occupied by white atoms be the concentration c. The fraction occupied by black atoms is then $1 - c$. At each value of c it is assumed black and white atoms occupy lattice sites randomly. We also assume for simplicity that the influence of atmospheric pressure on the thermodynamic stability of the system is negligible, which is usually the case. There is then no distinction between the Helmholtz and Gibbs free energies. It is possible these days to be quantitative and predictive for real alloy systems, avoiding many of the approximations and assumptions we are making here. The arguments below are intended to provide a qualitative understanding.

At a constant temperature T, the free energy per atom in each alloy may be expressed as $c\mu_W(c, T) + (1 - c)\mu_B(c, T)$, where $\mu_W(c, T)$ and $\mu_B(c, T)$ are the chemical potentials of white and black atoms in the alloy. The chemical potentials are functions of the concentration c and temperature T. The chemical potentials in the pure black and pure white substances are $\mu_B(0, T)$ and $\mu_W(1, T)$.

There are two contributions to the internal energy. The first is the heat content of the system. To a first approximation the heat content increases linearly with temperature. The second is the potential energy of atomic interactions. If the potential energy is independent of placing the black and white atoms on lattice sites randomly then the potential energy of the system varies linearly with c, as shown by the blue line in Fig. 2.2. This case is called an 'ideal solution'. If the potential energy is reduced by mixing black and white atoms then the internal energy varies as shown by the green line in Fig. 2.2. If the potential energy is reduced by separating black and white atoms into clusters of the same colour, the internal energy of occupying sites randomly with black and white atoms varies as shown by the red line in Fig. 2.2.

The other term in the free energy is $-TS$. The entropy has two contributions. The first arises from atomic vibrations. In this simple analysis it is assumed the vibrational entropy is independent of the concentration c, and it may therefore be ignored[2]. The

---

[2]There is also an electronic contribution to the entropy, which, it is assumed, is independent of c.

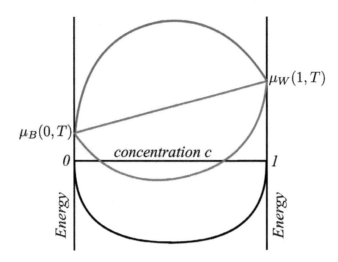

Fig. 2.2 Contributions to the free energy of a binary alloy of black and white components as a function of concentration c of the white component at a constant temperature T. Blue line: the internal energy per atom of a random alloy if there is no preference for mixing or clustering. Green curve: the internal energy per atom of a random alloy if mixing of atoms of different colours is favoured energetically. Red curve: the internal energy per atom of a random alloy if clustering of atoms of the same colour is favoured energetically. The end-points of the blue, green and red curves go up the energy axis with increasing temperature. Black curve: the $-TS$ term in the free energy per atom, where S is the configurational entropy of a random alloy. Notice that at $c = 0$ and $c = 1$ the black curve is vertical.

second is the configurational entropy of a random alloy which rises smoothly from zero in the pure systems to a maximum at $c = 0.5$, because the number of possible atomic configurations of the alloy is maximised when there are as many black atoms as white atoms. Therefore, at a given temperature the term $-TS$ varies with concentration c as shown by the black curve in Fig. 2.2.

The variation of the free energy, $A = U - TS$, with concentration c of the random alloys is obtained by 'adding' the black curve to each of the blue, green and red curves in Fig. 2.2. This produces the curves shown sketched in Fig. 2.3. These curves show the free energy per atom $c\mu_W(c, T) + (1 - c)\mu_B(c, T)$.

2.3   From free energy - composition curves to the equilibrium state

Consider the variation of the free energy with composition for an ideal binary solution, shown by the blue curve in Fig. 2.3. At each concentration c the minimum free energy is a random solid solution of black and white atoms. Not surprisingly, the same conclusion applies to the case where mixing black and white atoms is associated with a reduction of the potential energy, shown by the green curve in Fig. 2.3.

The case of the red curve in Fig. 2.3 is more interesting. It is reproduced in Fig. 2.4. The free energy of the random alloy with concentration c' is A_1. However, if the random alloy separates into B-rich and W-rich phases, with concentrations c_α and c_β

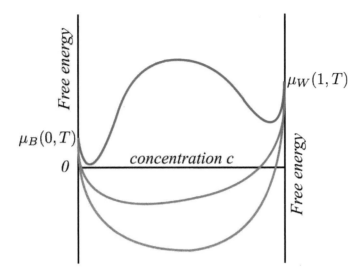

Fig. 2.3 The free energy per atom of a random solid solution as a function of the concentration c for a binary system obtained by adding, in turn, the black curve of Fig. 2.2, representing the $-TS$ term, to the blue, green and red curves of Fig. 2.2 representing three cases of the internal energy.

respectively, the free energy of the alloy with concentration c' decreases to A_2. The concentrations c_α and c_β are defined by where a common tangent touches the red curve. We call the phases with these concentrations α and β. Let the fraction of the alloy in the α-phase be f_α. The fraction in the β-phase is then $f_\beta = 1 - f_\alpha$. The total number of white atoms must be the same in the phase-separated alloy as in the random alloy. Therefore, $f_\alpha c_\alpha + f_\beta c_\beta = c'$. It follows that $f_\alpha = (c_\beta - c')/(c_\beta - c_\alpha)$ and $f_\beta = (c' - c_\alpha)/(c_\beta - c_\alpha)$. The free energy of the alloy with average composition c' is then $A_2 = f_\alpha A_\alpha + f_\beta A_\beta$, where A_α and A_β are the free energies of the α and β phases, as shown in Fig. 2.4.

It is obvious from the geometry of Fig. 2.4, and from the equation $A_2 = f_\alpha A_\alpha + f_\beta A_\beta$, that the minimum free energies of other alloys with concentrations between c_α and c_β are between A_α and A_β. The free energy per atom at $c = c_\alpha$ is $A_\alpha = c_\alpha \mu_W(c_\alpha, T) + (1 - c_\alpha)\mu_B(c_\alpha, T)$. The free energy per atom at $c = c_\beta$ is $A_\beta = c_\beta \mu_W(c_\beta, T) + (1 - c_\beta)\mu_B(c_\beta, T)$. The common tangent is a straight line and its equation is $A = A_\alpha + [(A_\beta - A_\alpha)/(c_\beta - c_\alpha)]c$, where the slope $[(A_\beta - A_\alpha)/(c_\beta - c_\alpha)]$ is a constant. Substituting the expressions for A_α and A_β into this equation we find the following relations must hold to ensure the slope is constant: $\mu_W(c_\alpha, T) = \mu_W(c_\beta, T)$ and $\mu_B(c_\alpha, T) = \mu_B(c_\beta, T)$. We recognise these relations as the condition for the coexistence of the α and β phases in thermodynamic equilibrium at constant temperature and pressure. The common tangent construction ensures the chemical potentials of black and white atoms are constant throughout the system after it has separated into two phases.

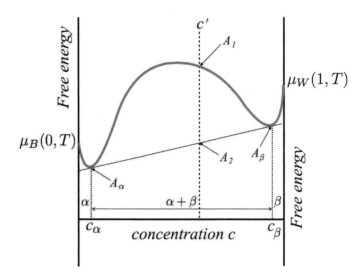

Fig. 2.4 The free energy of a random alloy with concentration c' is reduced from A_1 to A_2, which lies on the common tangent between A_α and A_β. The random solid solution with free energy A_1 decomposes into two phases with free energies A_α and A_β and concentrations c_α and c_β.

At concentrations between $c = 0$ and $c = c_\alpha$ only the α-phase is present. The α-phase is a solution of white atoms dispersed among primarily black atoms. The limit of the solubility is at $c = c_\alpha$. As the concentration of white atoms increases further the β-phase appears which is a solution of black atoms dispersed among primarily white atoms. The two-phase region exists between $c = c_\alpha$ and $c = c_\beta$. Between $c = c_\beta$ and $c = 1$ only the β-phase is present.

How does a structure like that shown in Fig. 1.1b arise? These structures occur at special (rational) atomic concentrations, such as $c = \frac{1}{4}, \frac{1}{2}, \frac{3}{4}$, where particular ordered crystal structures form. They are called 'intermediate phases' or 'ordered alloys'. They arise when the potential energy associated with chemical bonding in the crystal dominates over the entropy of mixing, strongly favouring particular atomic structures. Small deviations from these special concentrations raise the potential energy of the alloy very significantly. The dependence of the free energy on composition displays a sharp minimum at the special concentration, as illustrated in Fig. 2.5 for the case of an intermediate phase γ at $c = \frac{1}{2}$. Fig. 2.5 shows three phases α, β and γ. Notice that the γ-phase occurs by itself only over a very limited range of concentrations centred on $c = \frac{1}{2}$. It occurs also in the two-phase regions $\alpha + \gamma$ and $\gamma + \beta$. There is an example of an intermediate phase in Fig. 2.1. The hydrohalite phase with composition $NaCl \cdot 2H_2O$ appears at $c_{halite} = \frac{1}{3}$, which corresponds to a concentration of 61.9% by weight. The free energy composition curve rises so rapidly with deviations form the $c_{halite} = \frac{1}{3}$ composition the hydrohalite phase appears as a line in the phase diagram.

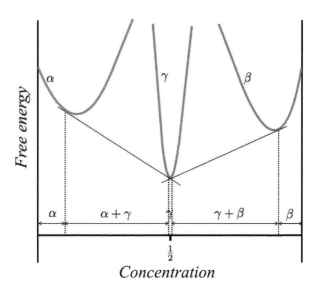

Concentration

Fig. 2.5 Free energy-composition curves in a binary alloy with an intermediate phase γ at $c = \frac{1}{2}$. Notice the narrow range of compositions where the γ-phase exists by itself and the adjacent two-phase regions $\alpha + \gamma$ and $\gamma + \beta$.

2.4 Phase diagram for complete miscibility

Fig. 2.6a shows the temperature-composition phase diagram at a given (e.g. atmospheric) pressure of two components B and W that are completely miscible in both the liquid and solid states. To be completely miscible in the solid state the components must share the same crystal structure and have similar atomic sizes and bonding. An example would be silver and gold. Figs. 2.6b-e show free energy-composition curves for the liquid and solid phases at a sequence of decreasing temperatures from T_1 to T_5. At the temperature T_1 the system is liquid at all compositions. As the temperature decreases the free energy of the liquid phase rises and the free energy of the solid phase falls. At T_2 pure component B solidifies. Pure component W solidifies at T_4. Between these temperatures liquid and solid phases coexist in the two-phase region $L+S$. According to the phase rule at a fixed pressure two phases may coexist in a two-component system with one degree of freedom. The degree of freedom is manifested in the range of temperatures between T_2 and T_4. When the temperature is below T_4, such as T_5, the alloy is a solid solution at all compositions.

The two-phase region is defined by the common tangent construction in Fig. 2.6d, where the solid phase has a smaller concentration of component W than the liquid with which it is in equilibrium, i.e. $c_1 < c_2$. With reference to Fig. 2.6a consider cooling the liquid alloy with concentration c_2. When the temperature reaches T_3 the first solid appears with a concentration c_1. On cooling through the two-phase region the concentration of the solid phase increases and eventually reaches c_2 when there is no liquid left. However, this assumes the cooling rate is sufficiently slow that redistribution of the component W in the solid phase can occur to maintain a uniform composition

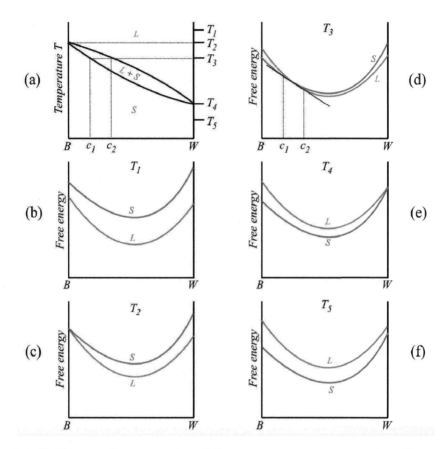

Fig. 2.6 To illustrate the construction of the temperature-composition phase diagram (a) for a two-component system where there is complete miscibility in the solid (S) and liquid (L) phases. The free energy-composition curves of the solid and liquid phases are shown in (b) to (f) at five temperatures T_1 to T_5.

at each temperature. In practice this does not usually happen and the solid phase usually has a concentration gradient along the growth direction.

2.5   Phase diagrams for limited solubility in the solid state

2.5.1   Eutectic phase diagram

When there is limited solubility in the solid state (apart from when there is an intermediate phase) more interesting phase diagrams arise. The eutectic we met in Fig. 2.1 is a common example. Its construction from five sets of free energy composition curves is illustrated in Fig. 2.7. Fig. 2.7e shows the free energies at the eutectic temperature where all three phases coexist. According to the phase rule three phases coexist at a unique temperature in a two component system when the pressure is specified. At the eutectic temperature, T_4, the liquid phase is transformed into the two solid

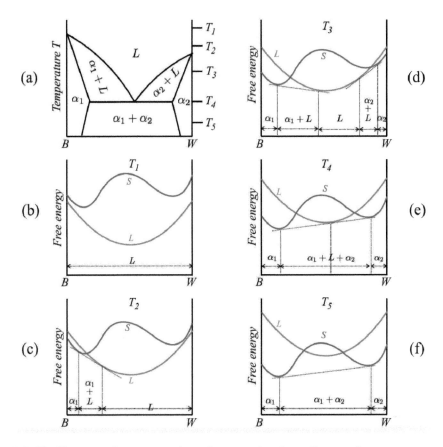

Fig. 2.7 To illustrate the construction of a eutectic phase diagram for a two-component system in (a) from free energy-composition curves at five temperatures in (b) to (f).

solutions α_1 and α_2. Note that the eutectic temperature is the lowest freezing temperature of all liquid compositions. It is common for the two solid phases to have different crystal structures. In that case they are represented by two independent free energy-composition curves, rather than the one (red) curve shown in Fig. 2.7b-f. As long as the free energy curve for the liquid comes between the minima of the free energy curves for the two solid phases a eutectic phase diagram will result.

2.5.2 Peritectic phase diagram

Another common binary phase diagram is the peritectic, shown in Fig. 2.8a. It also arises when there is limited solubility in the solid state. Examples of binary alloys displaying peritectic phase changes include platinum-rhenium and aluminium-titanium. In Fig. 2.8 the two solid phases, α and β, are independent, possibly with different crystal structures. This is not essential for the peritectic reaction, and they may be two concentration limits of the same phase as represented by the red curve in Fig. 2.7. In Fig. 2.8 the peritectic temperature is T_3. At this temperature the α and liquid

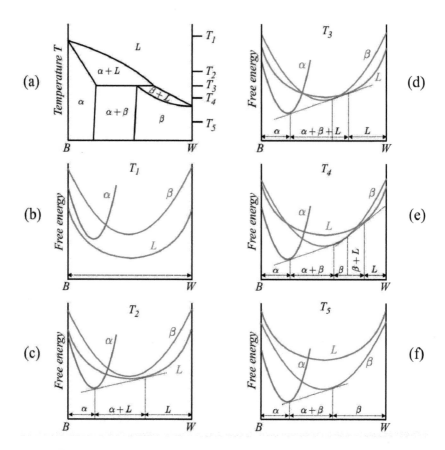

Fig. 2.8 To illustrate the construction of a peritectic phase diagram for a two-component system in (a) from free energy-composition curves at five temperatures in (b) to (f).

phases react to produce the β phase. Fig. 2.8e illustrates the free energy composition curves at the peritectic temperature. It is seen that there is a common tangent to the free energy curves of all three phases indicating they are in equilibrium at this unique temperature. As with the eutectic phase diagram, this is in accord with the phase rule.

2.6   Closing remarks

Phase diagrams may also be constructed for alloys of three components (ternary alloys) and more. There are no new principles involved but they are understandably more complex to construct and to visualise. The phase rule always applies.

We have stressed that thermodynamics and phase diagrams tell us nothing about the rate at which materials approach equilibrium. For that we have to consider the kinetics of the atomic mechanisms which bring about change in materials. In the next chapter we take a look at what is happening at the atomic scale in materials and some of the mechanisms by which change occurs.

Further reading

Cottrell, A H, *Theoretical structural metallurgy*, 2nd edition, Edward Arnold Ltd. (1960).

Denbigh, K, *The principles of chemical equilibrium*, Cambridge University Press (1964).

Gaskell, D R and Laughlin, D E, *Introduction to the Thermodynamics of Materials*, Taylor and Francis (2018).

Hillert, M, *Phase Equilibria Phase Diagrams and Phase Transformations*, Cambridge University Press (1998).

3
Restless motion

*In connection with the atomic structure of matter, it must be assumed
that the smallest components of matter, that is, atoms and molecules,
are not at rest but are in motion. Of course, this motion cannot be
seen directly; however, its existence is apparent from Brownian motion
...Heat energy is identical with the kinetic energy of the molecules
or atoms. Therefore, one can say that heat is a disordered form of energy.*

Reprinted by permission of The MIT Press from *Thermodynamics and
the kinetic theory of gases* by Wolfgang Pauli, edited by Charles P
Enz, translated by S Margulies and H R Lewis, foreword by Victor E
Weisskopf, Volume 3 of Pauli Lectures on Physics, p. 94. Copyright ©
1973 Massachusetts Institute of Technology.

3.1 Concept

Atoms in a solid are constantly moving. Many processes in materials are enabled by
transitory, local fluctuations of energy. The same fluctuations are also responsible for
reducing the mobility of defects.

3.2 Evidence of restless atomic motion

From the 18th century until the middle of the 19th century heat in a body was thought
to be associated with a weightless fluid called caloric, which flowed from hot to cold
bodies. In 1849 James Joule demonstrated experimentally that heat is a form of energy, equivalent to mechanical work. Joule showed that 3.0860 foot-pounds (i.e. 4.1858
Joules in SI units) of mechanical work is equivalent to 1 calorie of heat. A calorie is the
amount of heat required to raise the temperature of 1 gram of water by 1°C. These
days we tend to think of this equivalence as merely a conversion of units between
calories and Joules. But it was a huge step forward conceptually. Those who believed
in the existence of atoms soon recognised that heat was associated with the kinetic
energy of their motion relative to each other. The recognition that heat is a form of
energy was essential to the development of the conservation of energy and the first law
of thermodynamics.

If the atoms of a solid were completely fixed in position it would not be possible
for sound waves to be transmitted through it for the same reason that no sound can
be transmitted through a vacuum: sound waves require the movement of atoms. In
a metal heat is conducted primarily by free electrons, but in an insulator there are

virtually no free electrons. Therefore, it would also not be possible to conduct heat through a solid electrical insulator with immobile atoms. The electrical conductivity of a metal decreases with increasing temperature because a free electron travels a shorter distance before it is deflected by moving atoms. Such deflection of electrons is called scattering. The heat capacity of a body is the amount of heat required to raise its temperature by 1°C. If atoms in the body were in fixed positions they could not absorb energy, although in a metal the free electrons could still absorb energy. However, free electrons in a metal account for the observed temperature dependence of the heat capacity only at cryogenic temperatures. In insulators, and at all but the lowest temperatures in metals, the temperature dependence of the heat capacity is dominated by the energy absorbed by atomic vibrations.

3.3   Fluctuations and thermally activated processes

We saw in section 1.5.2 that when a macroscopic system is in thermodynamic equilibrium it does not mean nothing is changing at the atomic scale. At the atomic scale thermodynamic equilibrium is a dynamic state where the system samples the microstates available to it. For example, consider a macroscopic isolated system in which $\mathcal{R}$ is a small enclosed sub-system. The rest of the system behaves as a thermal reservoir with which $\mathcal{R}$ is in intimate contact. The internal energy of $\mathcal{R}$ fluctuates on the time scale of atomic vibrations as it exchanges energy with the rest of the system. Since the internal energy of the isolated system is strictly constant the fluctuations in the energy of the system outside $\mathcal{R}$ exactly compensate those inside $\mathcal{R}$. We will soon see that if $\mathcal{R}$ contains only 1-1000 atoms the fluctuations in its internal energy are relatively large in comparison to its time-average internal energy.

Spatially localised fluctuations of energy are enormously significant in materials science. They make possible a wide range of processes inside materials, including diffusion, aspects of permanent (plastic) deformation, transitions from brittle to ductile behaviour, softening through annealing, nucleation and growth of new phases and more. As we shall see in the next chapter, in crystalline materials all these processes involve the motion of crystal defects. There are energy barriers to the motion of defects. The energy required to surmount them is provided by local fluctuations. This is called thermal activation and the processes are said to be thermally activated. At lower temperatures the fluctuations from the average become smaller in magnitude. Consequently, the rates of thermally activated processes are reduced when the temperature is decreased. Although the fluctuations are larger at higher temperatures it should be clear that the potential energy component of the internal energy in $\mathcal{R}$ changes as well as the kinetic energy of atoms in the fluctuation. Consequently, the boundary of the region $\mathcal{R}$ fluctuates owing to variations in the forces between atoms inside and outside $\mathcal{R}$.

Fluctuations are a general feature of thermodynamic equilibrium, although thermodynamics is unable to account for them because they are transitory. The theory of fluctuations is the purview of statistical mechanics. But I can give a flavour. Let E and U be the instantaneous and time average values respectively of the internal energy of the sub-system $\mathcal{R}$. Let N be the number of atoms contained in $\mathcal{R}$. The time average value of the deviation of E from U is zero. That is because

$\langle E - U \rangle = \langle E \rangle - U = U - U = 0$, where $\langle X \rangle$ means the time average value of X. However, the time average value of the variance $(E - U)^2$ is not zero: $\langle (E - U)^2 \rangle = \langle E^2 - 2EU + U^2 \rangle = \langle E^2 \rangle - 2 \langle E \rangle U + U^2 = \langle E^2 \rangle - 2U^2 + U^2 = \langle E^2 \rangle - U^2$. In statistical mechanics it is shown that $\langle E^2 \rangle - U^2 = k_B T^2 N c_v$, where k_B is the Boltzmann constant, T is the temperature and c_v is the heat capacity per atom of the N atoms contained in $\mathcal{R}$. Therefore, $\langle (E - U)^2 \rangle / U^2 = k_B T^2 N c_v / U^2$. Since U is proportional to N the right-hand side is proportional to $1/N$. It follows that the fractional root mean square fluctuation $\sqrt{\langle (E - U)^2 \rangle}/U$ is proportional to $1/\sqrt{N}$. Thus, the magnitude of the fluctuations of the internal energy of the region $\mathcal{R}$, relative to its average internal energy, decreases as the size of $\mathcal{R}$ increases. In the limit that N becomes very large the fluctuations become negligible in relation to the time average value U. This is called 'the thermodynamic limit' because thermodynamics does not account for the existence of fluctuations. On the other hand, when N is very small the fluctuations become very significant. It is also evident that at absolute zero, where $T = 0 \, \mathrm{K} = -273.15°\mathrm{C}$, there are no fluctuations. But it turns out that even at this temperature atoms are still moving in a solid to satisfy the uncertainty principle of quantum mechanics. Atoms never stop moving in a solid, not even at absolute zero.

As an example of how fluctuations establish equilibrium, consider an isolated system comprising a solid in equilibrium with its vapour. Atoms in the vapour phase land on the surface of the solid phase. Some of them stick and some bounce off. At the same time fluctuations in the surface of the solid enable atoms to detach and enter the vapour phase. The surface is in a state of dynamic equilibrium with atoms arriving from the vapour and others departing into the vapour. Averaged over time there are as many attaching to the surface from the vapour as there are detaching from the surface and escaping into the vapour. This dynamic equilibrium is established through thermal fluctuations enabling atoms at the surface to escape the forces of attraction to their neighbours. If the temperature of the solid is raised suddenly, thermal fluctuations at the surface increase in size and frequency resulting initially in more atoms leaving the surface than joining it. The vapour pressure increases. Eventually the number of atoms in the vapour phase increases to the point where, averaged over time, as many join the surface as leave it. The system is then in equilibrium at the new higher temperature.

A considerable amount of additional energy has to be provided to an atom at the surface of the solid to enable it to enter the vapour phase. The additional energy is called an activation energy, and it is provided by a thermal fluctuation. For an atom to leave the surface its energy has to exceed the depth of the well of potential energy in which it sits. Not all atomic sites at the surface have the same well depth. Atoms in the shallowest well depths are the easiest to detach, and they are usually located at steps on a surface where atoms have the smallest number of neighbours.

A fundamental result of statistical mechanics is the probability p that the system is in an excited state, relative to the probability it is in the ground state. Thus, p is the ratio of two probabilities. At constant temperature and pressure we saw in the previous chapter that the ground state of the system is determined by minimisation of the Gibbs free energy. Let G_0 be this minimum Gibbs free energy. Let G be the Gibbs free energy of the system in an excited state. The ratio of the probability that the system is in the excited state to the probability it is in the ground state, is given by:

$$p = \frac{e^{-G/(k_B T)}}{e^{-G_0/(k_B T)}} = e^{-(G-G_0)/(k_B T)} \tag{3.1}$$

Since G_0 is the minimum Gibbs free energy $p \leq 1$, with $p = 1$ when $G = G_0$. This equation occurs throughout the theory of thermally activated processes. For a particularly clear derivation of eqn 3.1 see section 7.1 of the book by C P Flynn[1]. For an atom leaving the surface of the solid $G - G_0$ in eqn 3.1 is the activation free energy, which is the peak of the free energy barrier the atom has to overcome to escape the surface. It is important to recognise that G and G_0 are free energies, not merely potential energies, because the entropies of the ground and excited states also play a role.

3.4 Brownian motion

In Brownian motion a small particle suspended in a static liquid undergoes small random displacements as it is buffeted by molecules in the liquid. When it is very small the gravitational force on it is negligible in relation to the impulses it receives from the liquid. The average force the particle experiences is zero, but the fluctuations in the force are sufficient to displace the particle visibly under an optical microscope. The particle is an inert marker revealing the molecular fluctuations in the surrounding liquid. With a larger particle the random molecular buffeting is against its larger mass. Consequently, the particle moves less as its size increases until gravity takes over and it sinks to the bottom of the container, or rises to the surface, depending on its density and that of the liquid.

Brownian motion also occurs inside solids. It is the mechanism by which atoms diffuse in solids. Consider an interstitial atom in a crystal at equilibrium, such as carbon in iron. An interstitial atom occupies a space between the atoms of the crystal structure. These spaces are called interstices. The interstitial atom diffuses by jumping from one interstice to a neighbouring interstice. But for the jump to take place the host atoms, iron in the example, have to give the interstitial carbon atom a kick towards a neighbouring interstitial site. The host atoms are also temporarily in displaced positions allowing the interstitial to squeeze between them. These events occur at thermodynamic equilibrium through random, local fluctuations which supply the required local forces and energy.

Once the interstitial has jumped to a neighbouring site the energy supplied during the thermal fluctuation is dispersed into atomic vibrations that propagate away from the site of the jump. For the interstitial atom to make a second jump another thermal fluctuation is required, and the process repeats. Diffusion of the interstitial is a random walk through the interstitial sites of the crystal, as illustrated in Fig. 3.1. It is a thermally activated process. The random walk of the interstitial is a feature of the dynamical equilibrium state of the system as it samples the accessible microstates available to it through equilibrium fluctuations. Both the interstitial atom and the small particle suspended in a liquid are jostled by surrounding atoms and migrate along a random walk as a result. In both cases the mechanism of their migration is Brownian motion.

[1] C P Flynn, *Point Defects and Diffusion*, Clarendon Press (1972).

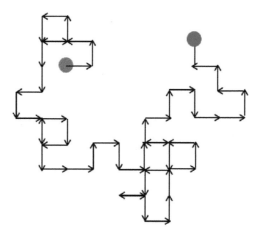

Fig. 3.1 A random walk on a square lattice. The walk starts at the green circle and finishes at the red circle. Notice that the distance between the start and finish is much less than the length of the random walk between them.

The distance an interstitial atom travels when it performs a random walk in a time t is approximately $\sqrt{Dt}$, where D is called the diffusivity. The diffusivity has units $m^2 s^{-1}$. For carbon in iron at $300°C$ the diffusivity is approximately $4.2 \times 10^{-14} m^2 s^{-1}$, so that in one second it travels about $200\,nm$ from where it started, but the path it took is much longer. At $600°C$ the diffusivity is approximately $1.8 \times 10^{-11} m^2 s^{-1}$, so that in one second it travels about $4000\,nm$ from where it started and again the path it took is much longer.

3.5 The fluctuation-dissipation theorem

Continuing with the example of carbon interstitials in iron, suppose there are cementite particles in the system where the chemical potential of carbon is less than it is in the iron matrix. Cementite is a separate phase with composition Fe_3C. The system is no longer in equilibrium because there are now gradients of the chemical potential of carbon atoms. These gradients generate forces on carbon interstitials. They are called driving forces because they drive change in the material. The driving forces are usually extremely small and in practice it means there is only a slight bias in the randomness of the jumps a carbon interstitial makes in favour of jumps towards a cementite particle. In this way the system approaches equilibrium by the carbon interstitials performing slightly biased random walks as they drift towards cementite particles. The energetics of a carbon atom jumping between two interstitial sites in the presence of a driving force are illustrated in Fig. 3.2.

Normally when we consider a force acting on a particle we think of Newton's second law, $F = ma$, with the force causing an acceleration. But as soon as the interstitial jumps, its energy is dissipated and it has to wait for another kick from a thermal fluctuation before it can move again. Another way of thinking about this is to say

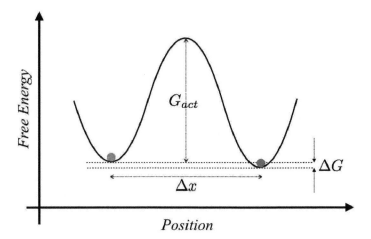

Fig. 3.2 Schematic illustration of the variation of the free energy of the system when a carbon atom migrates from the position represented by the green circle to a neighbouring interstitial site represented by the red circle in the presence of a driving force. $\Delta G < 0$ is the chemical potential of the interstitial at the red site minus that at the green site. $\Delta x > 0$ is the separation of the green and red sites. The driving force on the interstitial to move to the right is $-\Delta G/\Delta x > 0$. G_{act} is the activation free energy for the jump to the right: it is the free energy barrier the interstitial has to surmount. For the jump from red site to green site the barrier is larger by ΔG. In practice ΔG is much smaller than G_{act} than is depicted here. When $\Delta G = 0$, jumps between the red and green sites are equally probable.

that the interstitial experiences a 'drag force' which matches the driving force F, and the interstitial then moves at a constant average speed. As a result its drift towards a cementite particle is not characterised by an acceleration but an average speed. For small driving forces the average speed v of the interstitial is proportional to the driving force F, that is $v = MF$, where M is called the mobility of the interstitial. The units of M are $\mathrm{m\,s^{-1}N^{-1}} = \mathrm{s\,kg^{-1}}$.

There is a remarkable result, first discovered by Einstein in 1905, which shows there is a simple relationship between the diffusivity D of the interstitial and its mobility M, namely $D = k_B T M$. It is called the 'Einstein relation'. It is a surprising result because it relates an equilibrium property of the system, namely the diffusivity of the interstitial, to how the system responds when it is out of equilibrium, namely the mobility of the interstitial. It means that the drag forces on the interstitial in the presence of a driving force are directly related to the equilibrium fluctuations that enable it to perform a random walk in the absence of a driving force.

It may be helpful to consider an analogy. Suppose there is a room densely filled with energetic dancers. One of the dancers is very drunk. The dancers are analogous to atoms in the crystal and the drunk is analogous to the interstitial. The drunk is able to stand and stagger without falling over when a dancer bumps into him, but that is all. The position of the drunk changes only when he is struck sufficiently hard by one of the dancers. His movement around the dance floor is a 'random walk', and it is analogous to diffusion. But after a while, when he has sobered up a little, he spots some full bottles on the other side of the room. With a determined effort he tries to walk towards them, but he is buffeted by the dancers. The dancers are now reducing his ability to reach the bottles. Effectively they are exerting a drag force on him. Thus the dancers are responsible both for his diffusion in the absence of a driving force (visible bottles) and for limiting his mobility in the presence of a driving force.

Using the Einstein relation, and the carbon diffusivities in iron quoted at the end of the previous section, we may calculate the mobility of carbon atoms in iron at $300°C$ and $600°C$. We obtain $5.3 \times 10^6 \, \mathrm{s \, kg^{-1}}$ and $1.5 \times 10^9 \, \mathrm{s \, kg^{-1}}$ respectively. With a chemical potential difference of order $10^{-21} \, \mathrm{J}$ per atom over a distance of $1 \, \mu\mathrm{m}$ the average driving force is of order $10^{-15} \, \mathrm{N}$. We obtain drift speeds of order $5 \, \mathrm{nm \, s^{-1}}$ and $1 \, \mu\mathrm{m \, s^{-1}}$ respectively. Typical interatomic forces in metals are of order nanonewtons ($10^{-9} \, \mathrm{N}$). In this example the driving force on the carbon atoms is of order 10^{-6} times less than a typical force acting between atoms in the solid state. The driving force provides only a tiny bias in the direction of the jumps made by the interstitial.

The Einstein relation is an example of a general result in statistical mechanics called the fluctuation-dissipation theorem. Another example is the electrical resistance of a metal caused by thermal motion of atoms. Conduction electrons in a metal are accelerated by an applied electric field until they are scattered by vibrating atoms. In the scattering process the electrons transfer some of their kinetic energy to the atoms. Consequently, those atoms vibrate more and the metal heats up. Each conduction electron is repeatedly accelerated and scattered in various directions. Its average speed along the wire, which is called its drift speed, is much less than its speed between scattering events. This is one of the principal origins of electrical resistance in a metal. It explains why the resistance of a metal increases with temperature because then the atoms are vibrating with larger amplitudes and the frequency of scattering events increases. On the other hand, in the absence of an applied electric field thermal fluctuations in the circuit generate random electronic excitations and hence random voltages. These random voltages are the origin of 'thermal noise' in an electronic circuit. The resistance and the voltage fluctuations are related by a fluctuation-dissipation theorem. Thus the amount of thermal noise in a circuit increases with its electrical resistance.

An optical example is provided by the adsorption of light incident on a material. The photons of light lose their energy by exciting atomic vibrations and the material heats up. Thus, the energy of the light is dissipated into heat. On the other hand, when a material is heated electrons are excited to higher energy states by thermal fluctuations. Photons are emitted when the excited electrons fall from their excited states back to lower energy states. Thus, thermal fluctuations lead to the emission of photons. At normal temperatures the frequency of the emitted light is below the visible range in the far infra-red. But when materials are heated to high temperatures they become 'red hot' and then 'white hot'. The emission and absorption of photons are also related by a fluctuation-dissipation theorem. The more effectively a material absorbs light the more it will emit light when it is heated.

3.6   Some other manifestations of restless atomic motion in materials

A typical plastic carrier bag is made of low density polyethylene (LDPE). LDPE is a thermoplastic polymer, which means it melts on heating. It comprises long alkane molecules $(CH_3(CH_2)_n CH_3$, where $n \approx 10^4)$ forming an entangled mixture like a bowl of cooked spaghetti. In the liquid state fluctuations enable the long molecules to diffuse by a process called reptation. The word reptation comes from reptile. It is used because the molecules in liquid LDPE resemble snakes sliding past each other[2]. Each molecule in molten LDPE slides through a channel, defined by other molecules, when sufficient energy is provided by a fluctuation.

Another consequence of atomic motion is thermal expansion. Most solids expand when heated. In a crystal there are three unrelated reasons for this behaviour, two of which are rather subtle and will be discussed in section 4.3. The principal reason is that as the amplitudes of vibration of atoms about their equilibrium positions increase, the average separations of atoms also increase slightly. This is a result of a fundamental asymmetry in interatomic forces. Here we shall discuss qualitatively the thermal expansion of the bond in a diatomic molecule. The molecule can absorb energy in a variety of ways, including bond-stretching vibrations, rigid rotation and rigid motion of the molecule. We are concerned only with the absorption of energy at a higher ambient temperature by increasing the vibrational energy of the bond. For this discussion we assume classical physics rather than quantum physics.

As the atoms in the molecule are brought closer together from their equilibrium separation the potential energy rises steeply, as illustrated schematically in Fig. 3.3a. They repel each other and the force of repulsion increases as their separation decreases, see Fig. 3.3b. Part of the repulsion is quantum mechanical in nature arising from the Pauli exclusion principle (see section 6.5) and part of it is electrostatic arising from the overlap of the electron clouds.

If the separation of atoms is slightly larger than their equilibrium separation the increase of the potential energy is almost the same as when their separation is slightly less than the equilibrium separation. But as seen in Fig. 3.3a the potential energy rises more slowly as the atoms are separated further. When the atoms are far apart the potential energy of their interaction approaches zero as they become free of each other. The atoms initially experience a growing force of attraction with increasing separation, as shown in Fig. 3.3b. But unlike the force of repulsion, which continues to increase as atoms are brought closer together, the force of attraction reaches a maximum (negative value) and then decreases (becomes more positive) with further separation, eventually reaching zero when the atoms are free. The force of interaction between two atoms illustrated schematically in Fig. 3.3b is universal: at short range atoms repel, at long range they attract.

As the vibrational energy of the bond increases the range of separations between the atoms also increases. At a low vibrational energy the separation between the atoms might vary between points R_1 and R_2 in Fig. 3.3a. At a higher vibrational energy the

---

[2]Recall the scene in *Raiders of the Lost Ark* where Indiana Jones, played by Harrison Ford, is lowered into a pit of snakes writhing over each other.

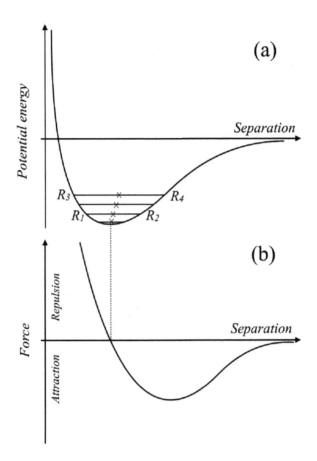

Fig. 3.3 (a) Potential energy of interaction between the atoms in a diatomic molecule as a function of their separation. The minimum in the curve corresponds to the equilibrium separation of atoms at absolute zero. R_1 and R_2 is the range of atomic separations at a relatively low ambient temperature. R_3 and R_4 is the range of atomic separations at a higher ambient temperature. The crosses are the mid-points of the ranges of separations. Notice that the mid-point is at a larger separation at the higher ambient temperature: the bond is longer on average. (b) shows the corresponding force acting between the atoms as a function of their separation. This is the force acting between the atoms at absolute zero. The dotted line shows that the force is zero at the minimum of the potential energy in (a).

separation between the atoms might vary between points R_3 and R_4 in Fig. 3.3a. We see that the mid-point of the range of separations increases slightly with increasing vibrational energy. In other words, the average bond length has increased slightly.

If the potential energy varied symmetrically on either side of the minimum, the mid-point of the range of separations between the atoms would not change with increasing vibrational energy, and the bond length would remain the same. This remains true even if the potential energy were not a parabola, as long as it is symmetric about the minimum. Although the argument is more complex in a solid, because each atom interacts with more than one other atom, the same conclusion is reached: thermal expansion is a macroscopic consequence of an asymmetry in the nature of interatomic forces. However, if the atoms of a solid were stationary there would be no thermal expansion.

If we ignore the zero point energy, which is the energy associated with zero point motion, the minimum of the potential energy of atomic interactions with respect to volume determines the separation of atoms in a solid at absolute zero. At a finite temperature and constant pressure it is the minimum of the *Gibbs free energy* of the solid with respect to volume that determines the equilibrium separation of atoms. The Gibbs free energy includes the internal energy and entropy of atomic vibrations as well as the potential energy of the atomic interactions. For a more complete treatment of thermal expansion in a solid see section 3.9 of the book by Sutton and Balluffi.

Further reading

Balluffi, R W, Allen, S M and Carter, W C, *Kinetics of Materials*, John Wiley (2005).

Cottrell, A H, *An Introduction to metallurgy*, 2nd edition, The Institute of Materials (1995).

Hinshelwood, C N, *The structure of physical chemistry*, Oxford University Press (2005).

Peierls, R E, *The laws of nature*, George Allen & Unwin Ltd. (1955).

Sethna, J P, *Entropy, order parameters and complexity*, Oxford University Press (2006).

Shewmon, P, *Diffusion in solids*, 2nd edition, The Minerals, Metals and Materials Society (1989).

Sutton, A P and Balluffi, R W, *Interfaces in crystalline materials*, Oxford University Press (2006).

Tabor, D, *Gases, liquids and solids*, Penguin Books (1969).

4
Defects

The dislocation is an object worthy of study. Its existence permits metals to be plastically deformed with ease, a circumstance upon which our modern technology is so dependent... The dislocation also permits nonmetallic crystalline materials to be plastically deformed... Thus the dislocation plays a commanding role in those grandest of all deformations on earth: the upheavals that have produced the mountain ranges and the continents themselves. Dislocations within grains of ice permit high mountains and high-latitude land masses to rid themselves, through the plastic flow of glaciers and ice sheets, of their ever-accumulating blanket of snow.

Reprinted with permission from Weertman, J and Weertman, J R, *Elementary Dislocation Theory*, Copyright © by Oxford University Press 1992, Inc.

4.1 Concept

Defects are the agents of change in crystalline materials.

4.2 Change in materials

The passage to thermodynamic equilibrium of a material with its environment may involve the formation of new phases. If the new phases involve changes of local chemical composition atoms have to be transported within the material by diffusion. Corrosion involves chemical reactions between a material and its environment. For example, oxidation resulting in the growth of an oxide layer on the surface of a metal occurs either at the surface of the oxide or at the surface of the metal or both. If it occurs at the surface of the oxide, metal cations enter and diffuse through the oxide layer to the surface where they combine with oxygen anions. If it occurs at the interface between the metal and the oxide layer oxygen anions enter and diffuse through the oxide layer to the metal where they combine with metal cations. Which of these occurs depends on a variety of factors including temperature and the partial pressure of oxygen in the gas phase. We shall see in this chapter that diffusion occurs through the migration of point defects in a crystalline material.

Some changes in materials are brought about deliberately. Industrial manufacturing turns raw materials into products. Metals are pressed, forged, extruded, drawn, milled, swaged, drilled and so on to create objects we use in our everyday lives. These processes all involve plastic deformation which we shall see involves the generation and movement of linear defects called dislocations. Ceramic products are insufficiently malleable to be shaped by the same manufacturing processes as metallic products. Instead they are usually made by sintering powders in preformed shapes at high temperatures and sometimes under high pressures. Sintering involves diffusion and it may also involve phase changes, glass transitions, and plastic deformation. It occurs through the generation and movement of point defects and dislocations.

There are also planar defects. Most crystalline materials are not single crystals. Instead they are agglomerates of many crystals called polycrystals. Each crystal or 'grain' is distinguished by the orientation of its crystal axes. Grain boundaries are the interfaces where grains of different orientations meet. They play a central role in mechanical, electrical and transport properties of polycrystalline materials. Some phase changes in materials occur through the rapid, diffusionless movement of interfaces which transform one crystal structure into another. The interfaces are then the agents of such transformations.

It is only a mild exaggeration to say that the science of crystalline materials is the study of defects and how their collective behaviour controls properties and processes in these materials. Even in some non-crystalline, glassy materials processes such as plastic deformation and the glass transition have been described in terms of the dynamics of defects defined as regions where local stresses exceed critical values.

The word 'defect' has pejorative connotations. But without defects we would not have many of the most useful properties of crystalline materials. Cracks are rarely beneficial, but to tar all defects with the same brush is to deny the essential role defects play in bringing about changes in materials our lives depend upon. Perhaps the more neutral term 'agent' is better because it connects the defect to a process. For example, dislocations are agents of plasticity, point defects are agents of diffusion, cracks are agents of fracture and so on. However, each type of defect is an agent of more than one process, but this only accentuates their central role in a diversity of properties and processes of crystalline materials.

4.3   Point defects

Imagine a perfect single crystal of pure copper at room temperature. Let us assume that every site of the crystal lattice is occupied by a copper atom. Suppose a thermal fluctuation results in an atom in the surface layer of the crystal jumping out of the layer to occupy a site on the surface, leaving behind an empty atomic site in the surface layer. The atom on the surface is called an adatom. The adatom may perform a random walk on the surface of the crystal, where there is a vast number of empty sites for it to occupy. Its random walk on the surface is an example of 'surface diffusion'.

The adatom can perform a random walk on the surface of the crystal because all the sites above the surface layer are available to receive the adatom. Atoms deep in the interior of the crystal are unable to move because all their neighbouring sites are occupied by other atoms. However, the vacant surface site the adatom left behind can move by a neighbouring atom jumping into it. If the neighbouring atom is beneath the surface layer the vacant site can begin to diffuse into the crystal interior. The vacant atomic site is called a vacancy and it is a point defect. An atom deep inside the crystal is able to move to an adjacent site only if the vacancy occupies the adjacent site first. When an atom jumps a vacancy makes the opposite jump. A net flux of atoms diffusing inside the crystal is effected by vacancies drifting in the opposite direction.

Other vacancies may enter the crystal by surface atoms jumping out of the surface layer and becoming adatoms. Thus a population of vacancies is established in the crystal. If enough surface atoms become adatoms whole new atomic layers are created at the surface as the vacancies diffuse into the crystal. In this way the number of crystal lattice sites that are either occupied by atoms or vacant increases. At a given temperature and pressure there is an equilibrium concentration of vacancies determined by minimisation of the Gibbs free energy of the crystal. At atmospheric pressure the Gibbs and Helmholtz free energies of copper are virtually identical. We may view the creation of a vacancy as the removal of an atom from an atomic site inside the crystal and its addition to the surface. Let the change in internal energy associated with this process be U_f. Let the change in the entropy of atomic vibrations associated with the process be S_f. The free energy associated with the creation of a single vacancy is thus $g_f = U_f - TS_f$. The free energy of the system includes the configurational entropy S_c associated with n vacancies on $N + n$ sites, where N is the number of atoms in the crystal. The configurational entropy is a property of the entire system and not any individual vacancy. The free energy of the system is thus $G = ng_f - TS_c$. At a finite temperature it is minimised when the fraction of sites occupied by a vacancy, $n/(N + n)$, is equal to $e^{-(g_f/k_B T)}$. Although the free energy cost g_f of creating vacancies is of order $1\,\text{eV}$ per vacancy ($\approx 100\,\text{kJ}\,\text{mol}^{-1}$) in metals, the configurational entropy of the system ensures there is a small but finite concentration of vacancies in thermodynamic equilibrium. In that case, to a good approximation, $n/(N + n) \approx n/N$. As usual, the equilibrium is dynamic with vacancies entering the crystal from the surface through the creation of adatoms at the same rate as other vacancies are annihilated by adatoms at the surface. Real crystal surfaces are rarely atomically flat. The generation and annihilation of vacancies takes place most easily at atomic steps on surfaces. They are also generated and annihilated at edge dislocations, described in the next section.

Diffusion by a vacancy mechanism involves both the formation *and* migration of vacancies. The activation free energy for diffusion by the vacancy mechanism is thus the sum of the free energies of formation and migration of vacancies. The activation free energy of self-diffusion can be determined by measuring, as a function of temperature, the penetration into a single crystal of radioactive isotopes of the element deposited on the surface of the crystal.

There are other conceivable mechanisms of diffusion inside the copper single crystal. An atom could leave its usual crystal lattice site and occupy an interstitial site. The atom then becomes a 'self-interstitial'. Diffusion would occur by the self-interstitial jumping to neighbouring interstitial sites. In contrast to the vacancy mechanism, where the number of crystal lattice sites available to be occupied by atoms is increased by the number of vacancies created, with the interstitial mechanism it is reduced by the number of self-interstitials. Except possibly at temperatures approaching the melting point, self-interstitials exist at equilibrium only in extremely small concentrations compared to those of vacancies because their internal energy of formation is generally so much higher than that of vacancies. However, interstitials may be created in large numbers by irradiation with high energy particles. Atoms are knocked off their crystal lattice sites by incident high energy particles and occupy interstitial sites, leaving behind as many vacancies. In general it is found that the activation energy of migration of these self-interstitials is significantly less than that of vacancies. The self-interstitials diffuse relatively rapidly and most, but not all, of them find vacancies within a few nanoseconds and recombine.

Another conceivable mechanism of self-diffusion is the direct exchange illustrated in Fig. 1.1d. However, in a single component crystal this does not lead to any diffusion because the atoms exchanging positions are indistinguishable, unless different isotopes of the same element are involved. The distortion required for two atoms to exchange positions is found to raise the internal energy to prohibitively high values. The distortion is less if a small group of atoms in a ring move collectively around the ring in the same direction. But again this does not lead to any diffusion unless atoms in the ring can be distinguished. In elemental metals the vacancy mechanism of diffusion is believed to be dominant[1].

Simmons and Balluffi performed a series of experiments in the 1960s on point defects in single crystals of aluminium, copper, silver and gold. These experiments established vacancies as the predominant point defects in thermodynamic equilibrium in these metals and measured their internal energies (U_f) and vibrational entropies (S_f) of formation. They are among the most elegant and definitive experiments in the history of materials science. A single crystal sample was heated over a range of temperatures to just below the melting point. At each temperature there were three contributions to changes in the volume of the crystal. The first was thermal expansion. Second, the point defects generated at each temperature had an elastic strain field which altered the average separation of adjacent atoms throughout the crystal. Third, the total number of crystal lattice sites that may be occupied or unoccupied by atoms changed. At each temperature the fractional change in the crystal lattice parameter $\Delta a/a$ was measured by X-ray diffraction, and the fractional change in the length $\Delta L/L$ of the specimen was measured with filar micrometer microscopes. Three times the fractional change in lattice parameter gives the first two contributions to the fractional change in the volume at a given temperature. At the same temperature, three times the fractional change in the length of the specimen gives all three contributions to the fractional change of the volume of the crystal. Therefore, $3(\Delta L/L - \Delta a/a)$ measures the change in the number of lattice sites in the crystal that can be occupied by atoms at each temperature. If vacancies are predominant then $3(\Delta L/L - \Delta a/a)$ is positive. If self-interstitials are predominant then $3(\Delta L/L - \Delta a/a)$ is negative. In all four metals vacancies are predominant, as illustrated in Fig. 4.1 for aluminium. By measuring the temperature dependence of $3(\Delta L/L - \Delta a/a)$ they were able to deduce the internal energy and vibrational entropy of formation of vacancies in all four metals. The results are summarised in Table 4.1.

By comparing the energies of formation of vacancies with the activation energies of diffusion it is found that the energies of migration of vacancies in these metals are

---

[1] It should be stressed that the mechanism of diffusion does not affect the discussion of Section 1.5.3 because the configurational entropy is independent of how the different configurations are achieved.

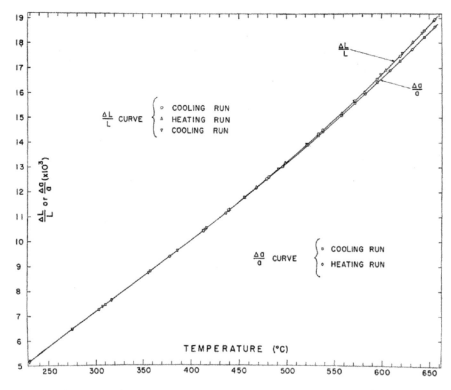

Fig. 4.1 $\Delta L/L$ and $\Delta a/a$ for a single crystal of aluminium plotted as a function of temperature. Reprinted figure with permission from Simmons, R O and Balluffi R W, Phys. Rev. **117**, 52 (1960). Copyright (1960) by the American Physical Society.

comparable to their energies of formation.

In an elemental insulator, such as pure silicon, essentially the same picture of diffusion holds, with the notable addition that the vacancies may now be electrically charged. There is an equilibrium population of vacancies for each charge state determined not only by the temperature but also by the position of the Fermi energy in the band gap[2]. Migration of charged point defects may require a compensating current of electrons or 'holes' to maintain charge neutrality.

So far we have considered point defects and diffusion in only single component systems. If there is a second component present at low concentrations it is often described as an impurity. Each impurity atom may also be regarded as a point defect. Considerable thermodynamic work has to be performed to overcome the configurational

---

[2] As explained in Chapter 6 the Fermi energy is determined by the energy of the highest occupied electronic state. Small concentrations of group V elements like phosphorus and group III elements like boron are deliberately introduced into silicon to position the Fermi energy near the top of the band gap or near the bottom of the band gap respectively. This is called *n*-type and *p*-type doping respectively. The position of the Fermi energy in the band gap determines whether electronic states associated with point defects in the band gap are occupied or unoccupied, and hence whether the defects are charged.

Metal	U_f(eV)	S_f/k_B
aluminium[a]	0.75 ± 0.07	2.4
copper[b]	1.17 ± 0.11	1.5 ± 0.5
silver[c]	1.09 ± 0.10	1.5 ± 0.5
gold[d]	0.94 ± 0.09	1.0

Table 4.1 Energy of formation, U_f, in eV and vibrational entropy of formation, S_f, expressed in units of the Boltzmann constant, k_B, of vacancies in single crystals of four metals, as measured by Simmons and Balluffi.
[a] Simmons, R O and Balluffi, R W, Phys. Rev. **117**, 52 (1960)
[b] Simmons, R O and Balluffi, R W, Phys. Rev. **129**, 1533 (1963)
[c] Simmons, R O and Balluffi, R W, Phys. Rev. **119**, 600 (1960)
[d] Simmons, R O and Balluffi, R W, Phys. Rev. **125**, 862 (1962)

entropy to remove impurities from crystals. Silicon-based technologies rely critically on controlling the impurity content at the level of parts per billion. The mechanism of diffusion of the impurity depends on whether it occupies a site that would otherwise be occupied by a host atom, or an interstitial site of the host crystal. The former is called a substitutional impurity, and the latter an interstitial impurity. Interstitial impurities tend to be small atoms compared to the host atoms and they diffuse by jumping between neighbouring interstitial sites. Substitutional impurities tend to have sizes comparable to or larger than those of the host and they normally diffuse by the same vacancy mechanism as the host atoms. In some cases vacancies become trapped by substitutional impurity atoms, particularly by large misfitting impurity atoms, forming 'complexes'. The mechanism of their migration can be considerably more complicated.

Point defects in ionic crystals have the additional complication that the atoms are charged because they are ionised. If there is a net flow of ionic charge in diffusion it can be sustained only if there is a compensating movement of electrons or holes[3] to maintain overall charge neutrality. Mass transport is then intimately coupled with transport of electrons and holes. Common salt is sodium chloride comprising Na^+ and Cl^- ions arranged on a cubic lattice. When a sodium ion is moved from inside the crystal to a site on the surface the Na^+ vacancy left behind has a formal charge of -1, in units of the absolute electronic charge, which is shared among the surrounding six Cl^- ions. To maintain local charge neutrality a vacancy on a chloride ion site may be created nearby by moving a chloride ion to the surface. The Cl^- vacancy has a formal charge of $+1$. Thus, the Na^+ and Cl^- vacancies attract each other electrostatically, and to separate them requires about 1.30 eV for each pair. Note that the chemical composition of the crystal is preserved by this pair of defects, which is called a Schottky defect. Another kind of defect that preserves the chemical composition is called the Frenkel defect. It comprises an ion leaving its usual site in the crystal lattice and

---

[3]In this context a 'hole' is a missing electron. It behaves like a vacancy in the electronic charge density with a charge of $+e$ where e is the magnitude of the charge on an electron. See also section 7.3.

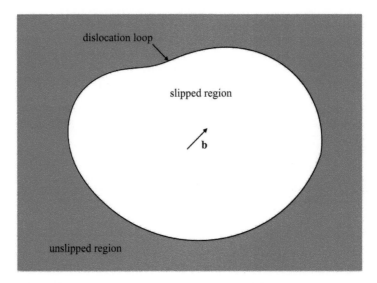

Fig. 4.2 A dislocation loop lying in a slip plane separating slipped and unslipped regions. The dislocation is the line separating the slipped and unslipped regions of the slip plane. It forms a loop. Inside the loop the material beneath the slip plane has been translated with respect to material above it by the Burgers vector **b**.

occupying an interstitial site. An example is the Ag^+ ion in silver chloride, which can occupy an interstitial site. In calcium fluoride, CaF_2, the fluoride ions F^- become Frenkel defects and occupy interstitial sites. The interstitial ion and the vacancy it leaves behind are oppositely charged and they attract each other electrostatically.

4.4   Dislocations

There are two principal ways in which solids are deformed: elastically and plastically. Elastic deformation is reversible. Plastic deformation is irreversible and permanent. A paper clip illustrates both. A straight metal wire is deformed plastically into the shape of a paper clip. When papers are inserted into the paper clip they are held in place by a small elastic deformation, which generates forces that pinch the papers holding them in place by friction. When the papers are removed the paper clip reverts to the shape it had before the papers were inserted and it can be used again.

Crystalline materials will deform elastically up to a certain point when they will either begin to deform plastically or they break. Understanding the transition from elastic to plastic deformation in the 1930s brought about the birth of the science of metallurgy, which broadened into materials science in the 1960s and 70s.

During plastic deformation of crystalline materials the volume of the crystal remains virtually constant. It occurs by a shearing process in which a plane of atoms slides over an adjacent atomic plane, like the shearing of a pack of cards. The sliding is called slip, and the geometrical plane midway between the two atomic planes undergoing slip is called the slip plane.

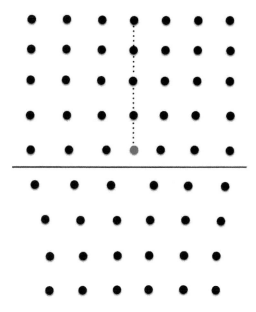

Fig. 4.3 A schematic illustration of an edge dislocation viewed along the dislocation line. Each dot represent a column of atoms normal to the page. The horizontal line is the trace of the slip plane. Above the slip plane there is an extra half plane shown by dotted lines. The edge dislocation is located at the termination of this extra half plane, which is shown by the red column of atoms.

The stress (i.e. force per unit area) required to shear an entire atomic plane *en masse* over an adjacent plane is several orders of magnitude higher than is observed experimentally in plastic deformation. This stress is called the theoretical shear strength of the material. Instead, slip begins in a patch of the plane, as illustrated in Fig. 4.2, and the patch grows until eventually the whole plane has undergone slip. The line separating the patch that has slipped from the rest of the slip plane that has not slipped is a dislocation. It forms a closed loop surrounding the slipped patch. The loop expands by the dislocation line moving normal to itself. As it expands the breaking and reforming of bonds across the slip plane is localised to the dislocation. It requires much less energy than breaking and reforming all the bonds across the slip plane simultaneously, as would be required if the plane slipped *en masse*.

The relative displacement of the atomic planes on either side of the slip plane has a constant magnitude and direction and it is called the Burgers vector. Provided the Burgers vector joins a pair of crystal lattice sites the crystal structures in the slipped and unslipped regions are the same. In that case plastic deformation does not change the crystal structure. Because the relative displacement is constant throughout the slipped region inside the loop it does not vary with the direction of the dislocation line. Where the direction of the dislocation line is perpendicular to the Burgers vector the dislocation is said to have 'edge' character. A pure edge dislocation is a long straight dislocation where the dislocation line is everywhere perpendicular to the

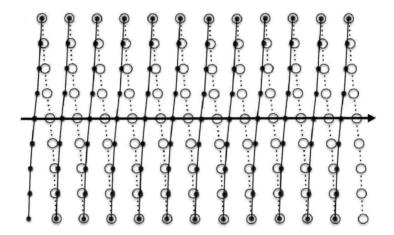

Fig. 4.4 A schematic illustration of a screw dislocation. The dislocation line is shown by the horizontal arrow. Solid circles and solid lines are underneath the page. Open circles and dotted lines are above the page. Planes of atoms are converted into a continuous spiral by the dislocation. This drawing was adapted from Hull and Bacon (2011).

Burgers vector. The structure of an edge dislocation is shown schematically in Fig. 4.3.

When the dislocation line is parallel to the Burgers vector the dislocation is said to have 'screw' character. A pure screw dislocation is a long straight dislocation where the dislocation line is everywhere parallel to the Burgers vector. The reason for its name is that the crystal lattice planes normal to the dislocation line become a helicoidal surface like the thread of a screw, as shown in Fig. 4.4. When the dislocation line is at some other angle to the Burgers vector it is said to have 'mixed' character, and its atomic structure is more difficult to visualise.

Suppose we see some defect in a crystal. How do we know it is a dislocation or some other defect? If it is a dislocation how do we determine its Burgers vector? The answer to both questions is the Burgers circuit construction. An example is shown in Fig. 5.5.

Gold is a soft ductile metal that is easily worked at room temperature to make jewellery. It deforms plastically very easily. Diamond is a very strong form of carbon that cannot be deformed plastically at room temperature. To make jewellery it can only be cleaved. Yet they are both crystalline and their crystal lattices are the same. Dislocations can exist in both materials, on the same slip planes with the same Burgers vectors. Why are their mechanical properties so different?

Carbon atoms in diamond are bonded to each other by strong covalent bonds along tetrahedral directions. The energy required to move the dislocation line *en masse* is too great because it involves breaking and making bonds all along the dislocation line. Instead it moves by the nucleation and propagation of kinks on the dislocation, as illustrated in Fig. 4.5. Bond breaking and making is now confined to the kinks, which are defects on the dislocation line and therefore they are point defects.

Dislocations localise slip to a line and kinks localise it further to points. Since kinks

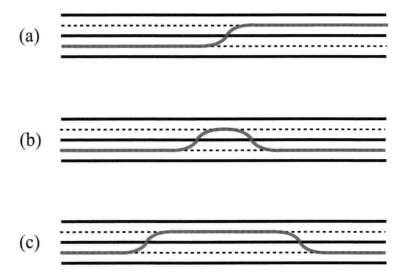

Fig. 4.5 Schematic illustrations of kinks. The slip plane is the plane of the paper. The dislocation line is shown in red. When the dislocation line is along the black solid lines its energy is a maximum. When it is along the broken lines its energy is a minimum. If the single kink in (a) moves to the left(right) the dislocation line moves up(down). For a straight dislocation line to move from one minimum to the next it has to bulge out as in (b), forming a 'double kink'. As the double kink (the bulge) expands in (c) more of the dislocation moves into the next minimum.

are point defects there is an equilibrium concentration of them on a straight dislocation at a given temperature. They also arise geometrically when the dislocation line changes direction as in a loop. The formation and migration of kinks are thermally activated events. In diamond the activation energies are high and they begin to occur sufficiently frequently to allow an observable rate of plastic deformation only at very high temperatures. At room temperature dislocations are virtually immobile in diamond, and it cleaves rather than deforming plastically.

Gold is different because its bonding is much less directional and much weaker. For example, the elastic shear modulus of diamond is around 500 GPa, whereas in gold it is only 27 GPa. As a result it is much easier to move dislocations in gold than it is in diamond. This comparison between gold and diamond shows that electronic structure and bonding are essential to understanding the mobilities of dislocations in different materials, even in materials sharing the same crystal lattices.

It is possible that dislocation lines in the same crystal and with the same Burgers vector have different mobilities when their line directions differ. For example, this is often the case in metals with body centred cubic crystal structures where screw dislocations are much less mobile than edge dislocations. This is a consequence of the different atomic structures of the cores of edge and screw dislocations in these metals, highlighting the role of the atomic scale in understanding the mobilities of dislocations in the same crystal.

Fig. 4.6 Grains visible to the naked eye in a zinc coating on a steel goal-post. The average grain size is about 2 cm.

Dislocations are the agents of plastic deformation in crystalline materials. Without them all crystalline materials would deform plastically only when the applied shear stress exceeds the theoretical shear strength. The material would probably fracture well before that stress is reached. It is difficult to imagine how different our world would be if metals were as brittle as diamond at room temperature.

The concept of dislocations also arises in earth sciences where it is used for example to describe earthquakes and the movement of glaciers. The study of dislocations is one of the many ways in which materials scientists and earth scientists share common interests.

4.5 Grain boundaries

Crystalline materials rarely exist as single crystals. As mentioned in Section 4.2 it is much more common for them to exist in a polycrystalline state consisting of many crystals, called grains, separated by interfaces called grain boundaries. The orientation of the crystal axes changes on crossing a grain boundary. The size of the grains varies from nanometres in nanocrystalline materials to centimetres. Fig. 4.6 shows grains in a galvanised zinc coating on a steel post.

Grain boundaries affect a wide range of properties and processes in crystalline materials. They are barriers to the motion of dislocations because the orientations of the slip planes and Burgers vectors change in the adjoining grain. Dislocations pile up against grain boundaries and form stress concentrations which may nucleate cracks or activate sources of dislocations in the next grain or in the grain boundary.

When the material deforms elastically the distribution of elastic stresses within the polycrystal is not uniform. The requirement that adjacent grains deform compatibly, that is without creating holes or overlapping material at the grain boundaries, produces additional stresses called compatibility stresses. They can be significant in magnitude and spatial extent in elastically anisotropic materials.

If a polycrystalline material is deformed plastically at a low temperature a fraction of the work done is stored in the material as the energy of defects generated during the deformation. These defects harden the material because they impede the motion of dislocations. If the material is subsequently annealed new grains are nucleated throughout the material and grow. The boundaries surrounding the new grains sweep through the deformed material, they absorb the defects created during plastic deformation, and they leave relatively defect-free material in their wake. This is called recrystallisation and it reduces the energy density associated with dislocations and other defects introduced by plastic deformation. It also softens the material. Grain boundaries are the agents of recrystallisation which is a restorative process in the material following plastic deformation. The growth of the new grains can be arrested by cooling the sample rapidly, and thus the grain size can be refined. Once recrystallisation has finished further grain growth may occur to reduce the energy density associated with the grain boundaries themselves, which further softens the material.

Grain boundaries sometimes act as channels for rapid transport of atoms, for example during the growth of oxide films on surfaces where ionic transport along grain boundaries may be much more significant than through the oxide grains.

Grain boundaries facilitate phase changes in the solid-state by acting as sites where the barrier to nucleation of new phases is reduced in comparison to that within the grains. This is called heterogeneous nucleation.

There are many other ways in which grain boundaries affect properties and processes in polycrystalline materials, including impurity segregation and embrittlement, sinks for point defects and point defect clusters in irradiated materials, Coble creep, grain boundary sliding as a mechanism of plastic deformation, and as sites of recombination of electrons and holes in polycrystalline photovoltaics.

Further reading

Cottrell, A H, *An Introduction to metallurgy*, 2nd edition, The Institute of Materials (1995).

Hull, D and Bacon, D J, *Introduction to dislocations*, 5th edition, Butterworth-Heinemann (2011).

Sutton, A P, *Physics of elasticity and crystal defects*, Oxford University Press (2020).

Sutton, A P and Balluffi, R W, *Interfaces in crystalline materials*, Oxford University Press (2006).

Weertman, J and Weertman, J R, *Elementary dislocation theory*, Oxford University Press (1992).

5
Symmetry

It is only slightly overstating the case to say that physics is the study of symmetry.

From Anderson, P W, *More is different*, Science **177**, 393 (1972). Reprinted with permission from AAAS.

5.1 Concept

Symmetry concerns invariance. It may refer to the invariance of an object to certain operations, such as rotations and translations. It may also refer to the invariance of equations governing motion of an object to operations on variables in the equations. Symmetry characterises physical properties of crystals. Topological defects, such as dislocations and disclinations, are characterised by translational and rotational symmetries of crystals.

5.2 Introduction

An object displays symmetry if its appearance is exactly the same after some operation has been performed on it. We say the object is invariant with respect to the operation. An equilateral triangle is invariant if it is rotated by 120° about a perpendicular axis through its centre, as shown in Fig. 5.1. It also has three axes of 180° rotational symmetry in the plane of the triangle. These are examples of rotational symmetries and they characterise the environment of the point at the centre of the triangle. An infinite crystal lattice displays translational symmetry. If the whole lattice is translated by any vector joining two lattice sites the lattice remains invariant. The enumeration of the various rotational and translational symmetries of crystals is the purview of crystallography, and it plays a central role in materials science. The symmetries of a crystal are discrete in the sense that the angles of the rotational symmetries may be only certain values and the translational symmetries are only lattice vectors. There is a finite number of rotational symmetries displayed by a crystal of infinite extent but there is an infinite number of discrete translational symmetries.

It may be surprising, but a gas or liquid of infinite extent also has symmetry when its atomic structure is averaged over time. Its time-averaged structure is invariant with respect to arbitrary translations and rotations. It has the same symmetry as empty space. It has an infinite number of rotational and translational symmetries, and they are described as continuous. When a gas or liquid becomes a crystal there is a reduction

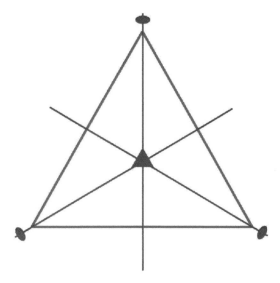

Fig. 5.1 An equilateral triangle (red) has a three-fold rotation axis normal to the page at the small blue triangle, and three two-fold rotation axes indicated by the blue lines and filled ellipses.

in the number of symmetry operations[1]. This is an example of 'broken symmetry'. However, there is an important sense in which the continuous symmetry operations that were present in the gas or liquid, and are absent in the crystal, still exist. The whole crystal can be translated and rotated by arbitrary amounts and remain invariant as far as all its physical properties are concerned. These arbitrary translations and rotations are the symmetry operations of the gas or liquid that are absent in the crystal. In other words, there is an infinite number of equivalent crystals that may be formed from the gas or liquid, which are related by the symmetry operations of the gas or liquid that are absent in the crystal, and which differ in their relative positions and orientations in space. That is how polycrystals, consisting of many crystals or 'grains' separated by grain boundaries, can exist. Each grain boundary is characterised by the change in orientation of the grains on either side of the boundary and the crystal planes that abut each other at the boundary. The existence of polycrystals is an example of the principle of symmetry compensation: *If symmetry is reduced at one structural level it arises and is preserved at another.*[2].

Another, perhaps less trivial, example of the principle of symmetry compensation is the phase change in barium titanate at around 120°C. Above 120°C the crystal structure is cubic and in the absence of an applied electric field it is not electrically polarised. Below 120°C the structure is tetragonal and the crystal has a spontaneous electric polarisation along the unique four-fold rotation axis. The unique four-fold

[1] The symmetry of a crystal also requires averaging over time to be seen because atoms are vibrating.
[2] A V Shubnikov and V A Koptsik, *Symmetry in science and art*, Plenum Press (1977), p.348

rotation axis of the tetragonal crystal is parallel to one of the four-fold rotation axes of the cubic crystal, of which there are three. There are two variants of the tetragonal crystal which may be obtained by applying to the tetragonal crystal the two four-fold rotations absent in the tetragonal crystal but present in the cubic crystal. Each variant has an electric polarisation along its four-fold rotation axis. To reduce the electrostatic and elastic energies of the crystal all three variants coexist below 120°C. Each variant occupies a domain. When an electric field is applied to the crystal below 120°C the domains that are aligned with the electric field grow at the expense of those that are not. Although the tetragonal crystal has less symmetry than the cubic crystal the existence of three variants of the tetragonal crystal is a consequence of the symmetry of the cubic crystal. In other words, the reduction in the symmetry of the crystal in going from a cubic crystal to a tetragonal crystal is compensated by the existence of three variants of the tetragonal crystal.

5.3   Conservation laws

A physical property of a system is conserved if it does not change no matter how the system evolves in time. The conservation of energy, of linear momentum and of angular momentum are examples of conservation laws in physics. It may be surprising that they are consequences of continuous symmetries displayed by the laws of physics. This follows from a theorem due to the German mathematician Emmy Noether in 1915. Her theorem states the following:

For every continuous symmetry of the laws of physics there is a conservation law.

She also proved the converse:

For every conservation law there is a continuous symmetry.

Noether's theorem is one of the most profound results in physics. Its influence on the development of many areas of physics has been huge – hence the quote by P W Anderson at the beginning of this chapter. The mathematics of Noether's theorem is beyond the scope of this book. Instead I shall offer two examples of familiar conservation laws that arise from continuous symmetries.

   Consider the example of a cluster of interacting particles in empty space where there are no forces acting on it from external fields. Let us suppose the particles interact with each other through some function describing their potential energy that depends only on their *relative* positions. If the position of every particle in the cluster is translated in space by the same amount their potential energy remains the same. Thus, the function describing their potential energy is invariant with respect to rigid translations of the cluster. The function that describes the total kinetic energies of the particles depends only on their masses and speeds. Therefore, this function also remains the same before and after a rigid translation of the entire cluster. There is a constant exchange in time between the *values* of the total potential and kinetic energies. Noether's theorem is not concerned with the values of the kinetic and potential energies but the symmetry displayed by the *functions* that determine them. These functions are invariant with respect to arbitrary rigid translations of the cluster. Therefore they display continuous symmetry with respect to translation. The presence of this continuous symmetry in Noether's theorem leads to conservation of the total momentum of the cluster.

The functions describing the total potential and total kinetic energies are also invariant with respect to translations of time. The presence of this continuous symmetry in Noether's theorem leads to conservation of the sum of the potential and kinetic energies of the cluster, i.e. conservation of its total energy.

As we have noted already the translational symmetry of a crystal lattice is discrete not continuous. This raises the question of whether the momentum of an electron moving in a perfect crystal (i.e. with no defects) is conserved, as it would be in empty space. For an electron moving in empty space its momentum is given by the de Broglie relation $p = hk$, where $k = 1/\lambda$ and λ is the de Broglie wavelength of the electron and h is the Planck constant. This momentum is the true momentum of the electron and it is conserved owing to the continuous translational symmetry of space. For an electron moving in a one-dimensional crystal lattice, where the lattice spacing is a, $p = hk = h/\lambda$ is conserved, but (i) p is no longer the true momentum of the electron and (ii) k can be replaced by $k + n/a$ where n is an integer. These two 'buts' are a consequence of the discrete translational symmetry of the crystal for which Noether's theorem does not apply. For an electron in a crystal the quantity hk is called the crystal momentum to distinguish it from the true momentum of the electron. In the language of quantum mechanics k is a good quantum number for an electron in a crystal. This means k provides a way to distinguish and label the states of electrons in a perfect crystal in the same way as s, p, d, f etc. label states with different angular momentum in the hydrogen atom.

5.4 Physical properties of crystals

Examples of physical properties of crystals include the electrical and thermal conductivities, the dielectric and magnetic susceptibilities, the diffusivity, the refractive index, the Young's modulus and the elastic stiffnesses and compliances more generally, the piezoelectric and thermoelectric coefficients and so on. In general these properties depend on the direction within a crystal along which they are measured.

For example consider the familiar example of the Young's modulus. A single crystal rod of length L and uniform cross-sectional area A is subjected to a tensile force F. The axis of the rod is along some direction in the crystal. The application of the force increases the length of the rod by ΔL. The Young's modulus Y relates the applied tensile stress F/A to the tensile strain $\Delta L/L$ as follows: $F/A = Y\Delta L/L$. The Young's modulus is a property of the crystal and it varies with the direction of the axis of the rod in the crystal lattice. Fig. 5.2 illustrates how the Young's modulus varies with direction in the crystal lattice in a single crystal of copper.

Notice in Fig. 5.2 that the variation of the Young's modulus with direction in the copper crystal has the symmetry of the crystal structure, which is also cubic. This is an example of a general rule known as Neumann's principle:

The symmetry elements of any physical property of a crystal must include all the symmetry elements displayed by the crystal.

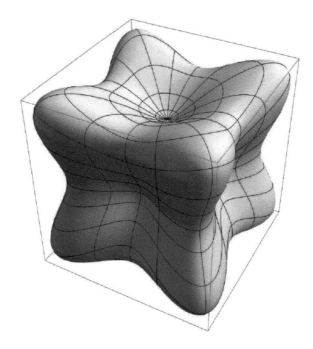

Fig. 5.2 A polar plot of the Young's modulus in a single crystal of copper. The bounding box is along the axes of the cubic crystal lattice. The Young's modulus is proportional to the length of the radius from the centre of the cube to a point on the coloured surface. The Young's modulus is a minimum (67 GPa) parallel to the cube edges and a maximum (191 GPa) parallel to the body diagonals. Notice that the plot of the Young's modulus has the symmetry of a cube.

In other words, when a property is measured along two directions related by a symmetry operation of the crystal the measurements must be identical. Although this principle does not tell us the value of the measured property it does tell us how many independent constants are required to characterise fully the physical property in different crystal structures.

In the absence of any rotational symmetry operations in a crystal there are 21 independent constants required to characterise fully its elastic properties. This is the case in a triclinic crystal[3] such as the mineral albite (NaAlSi$_3$O$_8$). The Young's modulus in any direction within a triclinic crystal is a function of all 21 independent elastic constants. Neumann's principle is unable to reduce this number because the crystal has no rotational symmetries. But in an orthorhombic crystal[4], with three perpendicular rotational axes of 180°, the number of independent elastic constants is reduced

---

[3] All three sides of the repeat cell of a triclinic crystal have different lengths and the angles between the faces are not 90°.

[4] The repeat cell of an orthorhombic crystal has a shape similar to a matchbox: the angles between the faces are 90° and the lengths of the three sides are different. The 180° rotational axes are normal to each face, and pass through each face centre.

by 12 to 9. Examples of orthorhombic crystals include olivine and cementite (Fe_3C). The former is the primary component of the Earth's upper mantle and the latter is a common phase in many steels. In cubic crystals there are just three independent elastic constants. In an isotropic material like rubber or glass the elastic properties are the same in all directions, and only two independent constants are required to characterise its elastic properties fully. For example, we could choose Young's modulus and Poisson's ratio. Poisson's ratio is the ratio of the lateral strain to the tensile strain of the rod. That is enough to determine how an isotropic material responds elastically to all kinds of loads including shear. The physical properties of isotropic materials have the symmetry of a sphere.

Neumann's principle contains the word 'include' because the symmetry of the physical property may be greater than that of the crystal. For example, the electrical and thermal conductivities and the diffusivity of cubic crystal structures have the symmetry of a sphere, which includes the symmetry of a cube.

5.5 Topological defects

A dislocation was introduced in Section 4.4 as a line defect in a crystal characterised by its Burgers vector and line direction. A dislocation loop encloses a region of the slip plane where the atomic planes on either side of the geometrical slip plane have undergone a relative displacement by the Burgers vector. The Burgers vector must be a translation vector of the crystal lattice if the crystal structure inside the loop is to remain the same as that outside the loop. Perfect dislocations have Burgers vectors that are lattice translation vectors. Inside the loop of an imperfect dislocation there is a planar fault in the crystal structure, which raises the internal energy of the crystal as the loop expands. The translational symmetry of the crystal lattice imposes restrictions on the possible Burgers vectors of perfect dislocations.

Dislocations are examples of topological defects. In contrast to point defects they cannot be removed by simply moving atoms around because they are associated with cuts in the medium. In Fig. 4.2 the medium inside the dislocation loop has undergone slip. Just below the slip plane inside the loop the medium is displaced by the Burgers vector relative the medium just above the slip plane. The displacement in the slip plane itself is not uniquely defined because it depends on whether we approach it from below or above. For this to be possible there must be a cut inside the loop on the slip plane. The creation of the dislocation loop may be envisaged as the following sequence of steps. First, the cut is made by switching off atomic bonds traversing the slip plane inside the loop. The relative displacement by the Burgers vector on either side of the cut is introduced. The bonding is switched back on. It is the presence of the cut and the jump in the displacement on either side of the cut that changes the topology of the medium and makes the dislocation a topological defect. For an isolated edge dislocation, such as that shown in Fig. 4.3, the cut extends from the dislocation to the edge of the crystal.

There are also topological line defects related to the rotational symmetry of the crystal. They are called disclinations. They are rare in 3D crystals because their energy is large. In 2D crystals such as graphene they are more common. Graphene is a sheet of graphite consisting of a hexagonal array of carbon atoms. Fig. 5.3a shows the perfect hexagonal pattern of graphene. In graphene there is a carbon atom at each vertex. There is an axis of six-fold rotational symmetry normal to the page passing through the centre of each hexagon. Suppose we remove the 60° wedge of atoms shaded green in Fig. 5.3a, pull together the two sides OA and OB of the remaining crystal and rebond so that all atoms have 3 bonds again. We would create the configuration shown in Fig. 5.3b where the red hexagon in Fig. 5.3a has become the red pentagon. This is a 60° wedge disclination. It has to have an angle that is consistent with the six-fold rotational symmetry of the lattice otherwise the two edges OA and OB will not be commensurate, and further defects will be created along them.

Alternatively after removing the 60° wedge of atoms in Fig. 5.3a we can force the two sides of the wedge further apart and insert a 120° wedge of atoms and rebond. The three-fold coordination of each carbon atom is again satisfied. In that case there is an additional 60° wedge of atoms inserted into the sheet and we create the configuration shown in Fig. 5.3c where the red hexagon of Fig. 5.3a has become the red heptagon. This is a 60° wedge disclination of the opposite sign to that shown in Fig. 5.3b. For both disclinations the displacement of atoms caused by the disclination is proportional to the distance from its centre. This is why they have such high energy.

By placing the two wedge disclinations of opposite sign adjacent to each other, as in Fig. 5.3(d), the growth in their long range displacement fields is cancelled and we create an edge dislocation. The terminating half planes of the edge dislocation are identified by the thicker lines in Fig. 5.4.

Fig. 5.5 illustrates the Burgers circuit construction for the edge dislocation shown in Figs. 5.3d and 5.4. Fig. 5.5a shows a right-handed circuit in green taken through lattice sites enclosing the dislocation. The circuit is closed so that it starts (S) and finishes (F) at the same site. In Fig. 5.5b the circuit is mapped onto a perfect hexagonal lattice where it no longer closes. The closure failure from the finish at F to the start at S is shown by the red arrow, and it defines the Burgers vector of the dislocation. The Burgers vector is a shortest lattice vector of the hexagonal lattice. Since the Burgers vector is a lattice vector the dislocation is perfect.

The 60° rotational symmetry of the hexagonal network has led to the identification of the heptagon and the pentagon as the elementary defects in the network. They are 60° wedge disclinations of opposite signs. By combining them, as in Fig. 5.3d we create a perfect edge dislocation. Other combinations of pentagons and heptagons in the hexagonal lattice create other defects including grain boundaries.

Disclinations involve making cuts to the edge of the sheet to remove or insert 60° wedges of material. The presence of these cuts makes disclinations topological defects. The formation of the dislocation in Fig. 5.3d involves making a cut to the edge of the sheet and inserting or removing a slice of material. This is illustrated in Fig. 5.6. Thus although they may appear to be point defects they are not like vacancies or impurities because they are associated with cuts that extend to the edges of the sheet.

5.6 Quasicrystals

It is a fundamental result in crystallography that the only rotational symmetries compatible with translational symmetry in a crystal are six-fold, four-fold, three-fold and two-fold rotations. Quasicrystals display long-range orientational symmetry but not

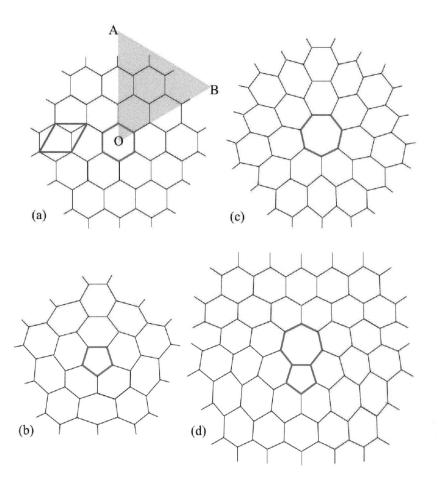

Fig. 5.3 (a) A perfect 2D hexagonal crystal. There is an atom at the vertex of each hexagon and each atom is bonded to three others. The blue rhombus is a repeat cell of the hexagonal lattice. It contains two atomic sites. The green shaded 60° wedge is cut out and removed in (b) and the two edges OA and OB of the remaining crystal are pulled together and rebonded to form a wedge disclination, with a pentagon at the centre shown in red. In (c) a disclination of opposite sign is produced by forcing a 120° wedge of atoms into the space created by removing the green shaded 60° wedge in (a). The red hexagon in (a) becomes the heptagon in (c). In (d) an edge dislocation is created by introducing wedge disclinations of opposite sign adjacent to each other.

translational symmetry. The absence of translational symmetry permits them to display different rotational symmetries, such as five-fold, ten-fold and twelve-fold rotations. The long-range order in the quasicrystal gives rise to discrete spots in its diffraction pattern, as opposed to the diffuse rings found in the diffraction pattern of metallic glasses.

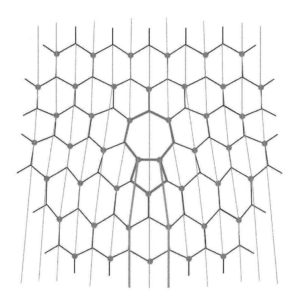

Fig. 5.4 The edge dislocation depicted in Fig. 5.3d showing the sites (blue) of the distorted lattice and the lattice lines in grey. Two thick grey lattice lines terminate at the bond shared by the heptagon and pentagon.

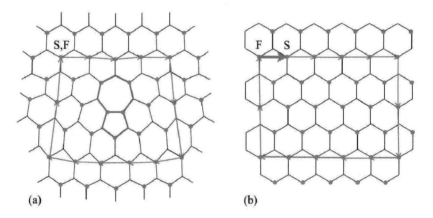

Fig. 5.5 To illustrate the construction of the Burgers circuit for the dislocation shown in Figs. 5.3d and 5.4.

The first quasicrystal discovered was an Al-14at% Mn alloy[5]. Its diffraction patterns demonstrated it had long-range icosahedral symmetry. Icosahedra appear frequently in intermetallic crystals, such as the Frank-Kasper phases, but they are arranged in periodically repeating cells that do not violate the above fundamental re-

---

[5]Shechtman, D, Blech, I, Gratias, D and Cahn, J W Phys. Rev. Lett. **53**, 1951 (1984). Shechtman was awarded the Nobel prize for chemistry in 2011 for making this discovery.

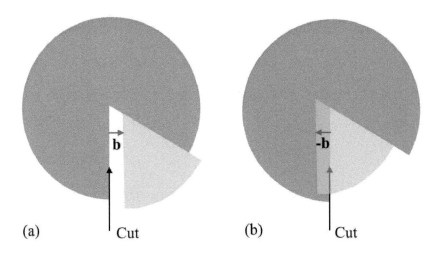

(a) 'Cut (b) 'Cut

Fig. 5.6 To illustrate the cut formed when two disclinations of opposite sign coexist to form an edge dislocation. In (a) material has to be inserted into the cut. In (b) overlapping material has to be removed. The Burgers vectors **b** of the edge dislocations formed in (a) and (b) have opposite signs.

sult of crystallography. There are no crystals with icosahedral symmetry. Since 1984 hundreds of intermetallic alloys have been found to be quasicrystalline. Some soft quasicrystals have also been found, for example in self-assembled colloidal micelles in solution. Quasicrystals are important conceptually because they are closely related to crystals in higher dimensions.

It may be puzzling that something can exist with a completely ordered structure without displaying translational symmetry. Consider a 2D square lattice of finite area. Suppose one edge of the lattice is along the direction (2,1). The edge comprises a periodic pattern of steps, like a staircase, with each step being two units wide and each riser one unit high. It has a slope of $\frac{1}{2}$. Suppose another edge of the lattice is along the direction (3,1). Again, there is a periodic pattern of steps, with each step now being three units wide. It has a slope of $\frac{1}{3}$. An intermediate edge orientation like (5,2) comprises a repetition of one (2,1) step followed by a (3,1) step. Similarly, a (7,3) edge is a repeating pattern of two (2,1) steps and a (3,1) step, and it has a slope of 3/7. These edges are along rational direction in the square lattice, that is they pass through sites of the square lattice, and the slope of each edge is a rational fraction.

The sequence of steps in any rational edge is easily determined by taking linear combinations of known sequences of other rational edges. But suppose we choose an irrational direction like $(e, 1)$ where e is the base of natural logarithms, $2.718281828459\ldots$, which is an irrational number. It will consist of a sequence of $(2, 1)$ and $(3, 1)$ steps, but this sequence never repeats exactly, no matter how large the square lattice.

Although the sequence of steps in the $(e, 1)$ edge never repeats it is completely determined. To see how we make increasingly accurate rational approximations to the $(e, 1)$ edge, starting from known repeat sequences with slopes above and below

$\frac{1}{e}$. The slope of $(e, 1)$ is between the slopes of $(11, 4)$ and $(8, 3)$. We choose $(11, 4)$ because it can only be a repeating sequence of three $(3, 1)$ steps and a $(2, 1)$ step. Similarly $(8, 3)$ can only be a repeating sequence of two $(3, 1)$ steps and a $(2, 1)$ step. Labelling a $(3, 1)$ step A and a $(2, 1)$ step B the repeat sequences of steps in $(11, 4)$ and $(8, 3)$ are $AAAB$ and AAB respectively. Combining these sequences we may form $(19, 7) = (11, 4) + (8, 3) = AAABAAB$. The slope of $(e, 1)$ is between the slopes of $(19, 7)$ and $(11, 4)$. If we form the sequence of $(68, 25)$ we are within 0.1% of the slope of $(e, 1)$. The repeat sequence of steps in $(68, 25)$ is found as follows:

$$(68, 25) = 3(19, 7) + (11, 4)$$
$$= 3(AAABAAB)AAAB$$
$$= AAABAABAAABAABAAABAABAAAB$$
$$= AAABAAABAABAAABAABAAABAAB$$

I used the periodicity of the entire sequence to move the sequence $AAAB$ at the end of the penultimate line to the beginning of the sequence in the last line. We can continue creating longer repeat sequences where the slope is ever closer to that of $(e, 1)$. In this way we can approximate the sequence of steps in the irrational edge $(e, 1)$ as accurately as we wish with rational approximants, each of which is completely determined. As the accuracy of the rational approximation improves the repeating sequence of steps becomes longer. In the irrational limit the sequence is infinitely long and it never repeats, but it is completely determined.

This simple example demonstrates how a completely ordered structure in one dimension – the sequence of steps along an edge – is determined by taking an irrational cut through a two-dimensional square lattice. The origin of the one-dimensional order in the sequence of steps in the irrational cut is the two-dimensional periodicity of the square lattice. The sequence of steps along the $(e, 1)$ edge is called quasiperiodic.

Mathematically, functions that display quasiperiodicity are easily created. The wave $\sin(2\pi x)$ has a wavelength of 1. The wave $\sin(\pi x)$ has a wavelength of 2. The wave $\sin(2x)$ has a wavelength of π. All three are periodic functions. The wave $\sin(2\pi x) + \sin(\pi x)$ is periodic with a wavelength of 2. The wave $\sin(2\pi x) + \sin(2x)$ is quasiperiodic. It never repeats exactly because the wavelengths of the constituent waves differ by a factor of π, which is an irrational number. The constituent waves are said to be incommensurate.

The 1D quasiperiodic function $\sin(2\pi x) + \sin(2x)$ can be generated from a 2D periodic function as follows. Consider $\sin(2\pi x) + \sin(2\pi y)$ which is a periodic function of x and y, as shown in Fig. 5.7. Let ξ be the position of a point with respect to the origin along the line $y = x/\pi$, shown in red. Then $\xi = \sqrt{x^2 + y^2} = \sqrt{1 + \pi^2}(x/\pi) = \sqrt{1 + \pi^2}y$. Therefore, $x = \pi\xi\sqrt{1 + \pi^2}$ and $y = \xi\sqrt{1 + \pi^2}$. Along the line $y = x/\pi$ the 2D periodic function $\sin(2\pi x) + \sin(2\pi y)$ becomes $\sin(2\pi x) + \sin(2x) = \sin(2\pi^2\xi\sqrt{1 + \pi^2}) + \sin(2\pi\xi\sqrt{1 + \pi^2})$. This function is plotted in Fig. 5.8.

It is interesting to examine the effect of shifting the origin of the 2D periodic function to $(d_x, 0)$ on the quasiperiodic function along $y = x/\pi$. The shift of origin is equivalent to a rigid translation of the 2D periodic function. The quasiperiodic function becomes $\sin\left[2\pi\left(\pi\xi\sqrt{1 + \pi^2} - d_x\right)\right] + \sin\left(2\pi\xi\sqrt{1 + \pi^2}\right)$. We see the effect

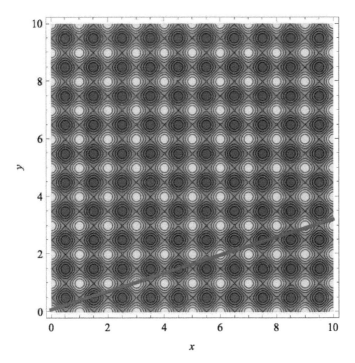

Fig. 5.7 Contour plot of the 2D periodic function $\sin(2\pi x) + \sin(2\pi y)$. The pale circles are minima and the dark circles are maxima. The red line is the line $y = x/\pi$. The value of the 2D periodic function along this line is plotted in Fig. 5.8.

is to introduce a phase difference between the two incommensurate waves along the x-axis. This phase difference is called a phason. It is not the same as shifting the origin of the quasiperiodic function, which merely produces a rigid displacement of the quasiperiodic function. The phason is a signature of the periodic function in the higher dimensional space.

The idea of generating a quasiperiodic structure from a periodic structure in a higher dimensional space carries over to real quasicrystals. Although five-fold rotational symmetry is not permissible with translation symmetry in three dimensions it turns out it can arise in a six-dimensional crystal. The quasicrystal may be thought of as an irrational three-dimensional hyperplane cut through the six-dimensional cubic crystal. Dislocations can and do exist in quasicrystals but with Burgers vectors defined by circuits in the six-dimensional crystal lattice. A dislocation in the higher dimensional crystal lattice brings about a spatially varying phason field as well as an elastic displacement field in a quasicrystal, and the effects of both may be detected in an electron microscope. The result is that the Burgers vector of a dislocation in a quasicrystal, like each of the spots in the diffraction pattern of a quasicrystal, has six independent components.

Although quasicrystals do not display translational symmetry they do display local isomorphism. This means that any finite region of an infinite quasicrystal is repeated

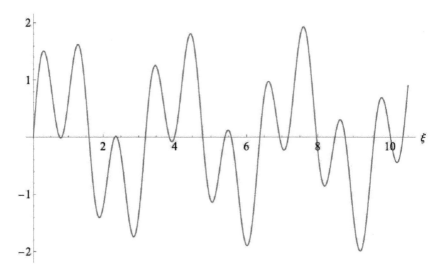

Fig. 5.8 The function depicted in Fig. 5.7 along the line $y = x/\pi$. ξ is the coordinate along the line. The function in this plot never repeats exactly. It is quasiperiodic.

infinitely many times, but the distance from the finite region before a repetition is found increases with the size of the repeating region. There are many other fascinating aspects of quasicrystals, including Penrose tilings and the role of the golden mean. Irrational interfaces between ordinary crystals may also be described as quasiperiodic structures[6]

Further reading

Janot, C, *Quasicrystals: a primer*, 2nd edition, Oxford University Press (2012).

Lederman, L M and Hill, C T, *Symmetry and the beautiful universe*, Prometheus Books (2008).

Newnham, R E, *Properties of materials*, Oxford University Press (2005).

Nye, J F, *Physical properties of crystals*, Oxford University Press (1985).

[6]Sutton, A P, Progress in Materials Science **36**, 167 (1992).

6
Quantum behaviour

It was terribly laborious – I didn't understand matrix calculus then . . . I get so excited I keep making mistakes. But by three in the morning I've got it. I seem to be looking through the surface of atomic phenomena into a strangely beautiful interior world. A world of pure mathematical structures. I'm too excited to sleep. I go down to the southern end of the island. There's a rock jutting out into the sea that I've been longing to climb. I get up it in the half-light before the dawn, and lie on top, gazing out to sea.

The character of Werner Heisenberg speaking about a crucial moment of discovery concerning quantum mechanics while he was alone on holiday on the island of Heligoland in the North Sea, in Act 2 of *Copenhagen* by Michael Frayn. First published by Methuen in 1998. © Michael Frayn. All rights reserved. Extract reprinted by kind permission of Michael Frayn, c/o United Agents.

Those who are not shocked when they first come across quantum theory cannot possibly have understood it.

Niels Bohr, quoted by Werner Heisenberg in *Physics and Beyond: Encounters and Conversations*, p.206. Harper & Row (1972).

I think I can safely say that nobody understands quantum mechanics.

Richard Feynman in *The Character of Physical Law*, p.129. Published by The British Broadcasting Corporation. Copyright © Richard Feynman 1965.

6.1   Concept

In a universe obeying classical physics only, matter as we know it would not exist. At the most fundamental level the structure and properties of all materials are governed by quantum physics.

6.2   The size and identity of atoms

Atoms have a finite size of the order of $1\,\text{Å}$, which is $10^{-10}\,\text{m}$. The atomic nucleus is much smaller, of order $10^{-15} - 10^{-14}\,\text{m}$. According to classical electrodynamics an accelerating charge emits electromagnetic radiation. Indeed, this is the mechanism of generating X-rays in a synchrotron where charged particles travel at high speeds around a circle. Until the development of quantum mechanics it was not clear why electrons orbiting an atomic nucleus did not collapse into the electrostatically attractive nucleus as a result of radiating energy in the form of electromagnetic waves.

When hydrogen atoms are heated they emit light as electrons in excited states fall into lower energy states. The emission spectrum of hydrogen consists of discrete lines at well defined, reproducible frequencies rather than a continuous spectrum. The spectrum of lines is so reproducible and unique to hydrogen it can be used to identify the existence of hydrogen anywhere in the Universe. The Bohr model of the hydrogen atom was able to reproduce the emission spectrum by *postulating* that the electron is a wave with a de Broglie wavelength λ that has to be commensurate with the circumference $2\pi r$ of the circular orbit of the electron: $2\pi r = n\lambda = nh/(mv)$ where $n = 1, 2, 3, \ldots$, h is the Planck constant, m is the electron mass and v its speed. In Bohr's model and in Newtonian mechanics the radius r and the speed v are related by equating the centripetal and electrostatic forces, $mv^2/r = e^2/(4\pi\epsilon_0 r^2)$, where e is the magnitude of the electron charge and ϵ_0 is the permittivity of free space. In Newtonian mechanics this relationship is the only constraint on r and v; they are otherwise arbitrary. Thus, in Newtonian mechanics the size of a hydrogen atom is not uniquely defined. Bohr's postulate fixes r and v for a given value of n. For $n = 1$ the radius of the electron orbit has the value $0.529\,\text{Å}$. The well defined value of r in the ground state $n = 1$ in Bohr's model was supported by experiments which showed that hydrogen atoms are identical to each other. Bohr's central postulate was a brilliant insight but it lacked theoretical justification. For example, his model asserted rather than explained that the electron in a hydrogen atom did not radiate electromagnetic waves or collapse into the nucleus.

The full resolution of these puzzles came with the development of the Schrödinger equation and its application to the hydrogen atom. The Schrödinger equation shows that the electron in a hydrogen atom has a well defined minimum energy state, where it is not possible for it to access lower energy states by radiating energy. The well defined nature of the ground-state configuration also ensures that all hydrogen atoms are identical in their ground states.

6.3 The double slit experiment

The double slit experiment by Thomas Young at the beginning of the nineteenth century showed that light has wave-like properties. In the modern version of this famous experiment light of a single wavelength passes through two closely spaced, parallel slits in a screen and impinges on an observation screen. An interference pattern of bright and dark lines is formed on the observation screen. Let $d_1(x)$ and $d_2(x)$ be the distance from each slit to a point x on the screen. When the path difference $d_1(x) - d_2(x)$ is a whole number of wavelengths of the light we get constructive interference and a bright region on the observation screen. When $d_1(x)$ and $d_2(x)$ differ by a whole number of wavelengths plus half a wavelength the light waves interfere destructively and we see a dark region on the observation screen. The pattern seen on the observation screen is thus due to the interference of waves of light emerging from the two slits.

If one slit is covered so that the light passes through only one slit the interference pattern disappears. The interference pattern is replaced by a smooth distribution of light intensity, which we call a single slit distribution.

I was given the above explanation of the Young double slit experiment at school fifty years ago. The explanation is exactly the same if instead of light we had a steady source of waves in a pond, which pass through two narrowly spaced gaps in a barrier on the surface of the water and impinge on an absorbing surface (so that the waves are not reflected). At the absorbing surface we see the same pattern of constructive and destructive interference.

But there is a lot more to the experiment with light. In 1905 Einstein explained the photoelectric effect, by showing that light consists of particles, called photons. When light is shone on a metal surface electrons are ejected from the surface provided the energy of the photon exceeds the binding energy of the electron to the metal. Each photon has energy $h\nu$ and momentum h/λ, where ν and λ are the frequency and wavelength of the light. It was his explanation of the photoelectric effect that earnt Einstein the Nobel Prize in physics in 1921.

Einstein's explanation of the photoelectric effect raises the question that if light is a particle how do we get the interference pattern in the double slit experiment? We might think a photon would have to travel through only one of the slits since, as a particle, it could not travel through both. It is conceivable there is interference between a photon travelling through one slit and another photon travelling through the other slit at the same time. If that were so then if we reduce the intensity of the light source, so that there is only one photon travelling through the apparatus at a time, the interference pattern should disappear. *But it does not disappear.* It takes longer for the interference pattern to appear on the observation screen but the same pattern eventually emerges.

A related experiment was conducted in 1909 by G I Taylor[1] when he was still a student in Cambridge working with J J Thomson who had discovered the electron in 1897[2]. This was Taylor's first paper. It was entitled *Interference fringes with feeble light*. The fringes were produced in the shadow of a needle on a photographic plate. The light source was a narrow slit placed in front of a gas flame. The intensity of the light was weakened by passing through a series of smoked glass plates. The experiment with the weakest illumination required an exposure time of three months. The result was that even the weakest light source produced the same interference pattern in the shadow of the needle.

Unfortunately, the gas flame used by Taylor can emit more than one photon at a time, and it could be argued that the pattern he observed was achieved by interference between more than one photon, even when the illumination of the needle was weakest. More recent experiments[3] with single photon sources have established that the inter-

---

[1]G I Taylor went on to make seminal contributions to materials science. In 1934 he published a particularly far-sighted paper proposing dislocations as the agents of plasticity and work-hardening in metals, long before dislocations were observed in metals. Despite his enormous contributions to materials science he is probably best known for his work in fluid dynamics and geophysics. He said he "did not feel a call to a career in pure physics".

[2]Taylor, G I, Proc. Camb. Phil. Soc. bf 15, 114-115 (1909).

[3]Grangier, P, Roger, G and Aspect, A, Europhys. Lett. **1**, 173 (1986).

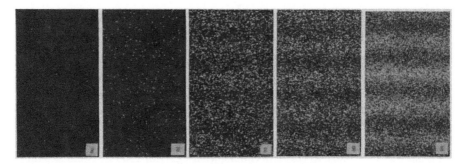

Fig. 6.1 The build up in time of the Young's double slit interference pattern by electrons in an electron microscope. In this experiment there was only one electron passing through the electron microscope at a time. Each white dot corresponds to the arrival of a single electron. From left to right the number of electrons in each frame is 10, 100, 3,000, 20,000 and 70,000. Reproduced from Tonomura *et al.* (1989)[5], with the permission of the American Association of Physics Teachers.

ference pattern seen in the double slit experiment is maintained when just one photon at a time passes through the slits. This cannot be explained by classical physics.

The double slit experiment with photons confirms Einstein's discovery that light sometimes behaves like a particle and sometimes like a wave. But what is its relevance to materials? Instead of using photons in the experiment we can use particles with mass. The experiment has been carried out by passing electrons one at a time through an electron biprism which behaves like a double slit. This was first done in 1974 by Merli *et al.*[4] using a thermionic electron source in an electron microscope. The experiment was repeated independently in 1989 by Tonomura *et al.*[5] using a field emission electron source in an electron microscope. Figure 6.1 shows how the interference pattern gradually emerges from single electrons passing through the electron biprism in the experiment by Tonomura *et al.* The interference pattern demonstrates that electrons behave as waves, while the white dots they create on the screen demonstrate they behaves as particles. Thus, electrons have the same wave-particle duality as photons.

Suppose we introduce a detector that enables us to determine which slit the electron passes through. This was not done in the experiments by Merli *et al.* and Tonomura *et al.* and it is a thought experiment. If we ran the experiment again we would find the interference pattern is replaced by a superposition of two single slit distributions, one from each slit. If we watch which slit each electron passes through we destroy the interference pattern! The explanation for this bizarre behaviour is that whatever the mechanism of the detector it will disturb the electrons. There will be an exchange of energy or momentum or both between the electron and the detector. Thus, by introducing a detector we are disturbing the delicate state of the electrons and this

[4]Merli, P G, Missiroli, G F and Pozzi, G, American Journal of Physics **44**, 306 (1976).

[5]Tonomura, A, Endo, J, Matsuda, T, Kawasaki, T and Ezawa, H, American Journal of Physics **57**, 117 (1989).

destroys the interference. This highlights a problem of measurement in the quantum world, namely the act of measurement disturbs the system significantly.

In summary, if we are able to detect which slit the electron passes through then we can say it passes either through slit 1 or slit 2. In that case there is no interference pattern, just a combination of the single slit distributions from both slits. But if we do not disturb the electrons by trying to detect them, we do not know which slit they pass through. Since we do not know, we may *not* say the electron passes through either slit 1 or slit 2. In that case we do see the interference pattern. When we observe the electrons we force them to pass through either slit 1 or slit 2. But when we don't observe them it is as if they are passing through both slits and interfering with themselves.

The mathematical expression of these experiments in quantum mechanics is in terms of probabilities and probability amplitudes. The probability P of an event is *defined* by $|A|^2$ where A is a complex number representing the probability amplitude for the event. This way of defining probability is unique to quantum mechanics, and the only justification for it is that it leads to agreement with experiment. Since A is a complex number it has a modulus $|A|$ and a phase ϕ, so that $A = |A|\,e^{i\phi}$.

Let $A_1(x) = |A_1(x)|\,e^{i\phi_1(x)}$ be the probability amplitude that an electron passes through the first slit and arrives at position x on the observation screen. When the second slit is covered we know all the electrons pass through the first slit. The probability that the electron arrives at x on the observation screen is then given by $P_1(x) = |A_1(x)|^2$. This is what we called a single slit distribution. Let $A_2(x) = |A_2(x)|\,e^{i\phi_2(x)}$ be the probability amplitude that an electron passes through the second slit and arrives at position x on the observation screen. If the first slit is covered then the probability $P_2(x)$ that the electron arrives at x on the observation screen is $|A_2(x)|^2$. If we uncover both slits and use a detector to tell us which slit the electron passes through then the probability of the electron arriving at x on the observation screen is just $P_{1 \text{ or } 2} = P_1(x) + P_2(x)$, and there is no interference.

If both slits are uncovered and we don't know which slit each electron passes through the probability amplitude is $A_{1 \text{ and } 2}(x) = A_1(x) + A_2(x)$. The probability of the electron arriving at x on the observation screen is then:

$$
\begin{aligned}
P_{1 \text{ and } 2}(x) &= |A_{1 \text{ and } 2}(x)|^2 \\
&= (A_1(x) + A_2(x))\,(A_1^\star(x) + A_2^\star(x)) \\
&= A_1(x)A_1^\star(x) + A_2(x)A_2^\star(x) + A_1(x)A_2^\star(x) + A_2(x)A_1^\star(x) \\
&= |A_1(x)|^2 + |A_2(x)|^2 + |A_1(x)|\,|A_2(x)| \left(e^{i(\phi_1-\phi_2)} + e^{i(\phi_2-\phi_1)} \right) \\
&= P_1(x) + P_2(x) + 2\,|A_1(x)|\,|A_2(x)| \cos\left[\phi_1(x) - \phi_2(x) \right] \\
&= P_{1 \text{ or } 2}(x) + 2\,|A_1(x)|\,|A_2(x)| \cos\left[\phi_1(x) - \phi_2(x) \right].
\end{aligned}
\tag{6.1}
$$

The star denotes complex conjugation, i.e. $A_1^\star(x) = |A_1(x)|\,e^{-i\phi_1(x)}$ and $A_2^\star(x) = |A_2(x)|\,e^{-i\phi_2(x)}$. The interference term $2\,|A_1(x)|\,|A_2(x)| \cos\left[\phi_1(x) - \phi_2(x)\right]$ gives rise to the bright and dark fringes at the observation screen.

The experiment has been carried out with more massive particles including neutrons, atoms and even molecules. For example, C_{60} molecules (bucky balls) have been shown to display wave-particle duality[6]. The de Broglie wavelength of the particle is inversely proportional to its mass. The de Broglie wavelength of the C_{60} particles in this experiment was around 2.5×10^{-12} m. The de Broglie wavelength for bullets fired from a machine gun is of order 10^{-35} m. If we fired bullets from a machine gun at a double slit we would not see an interference pattern because the spacing of the fringes would be far too small to resolve. We would see only the envelope function $P_{1 \text{ or } 2}(x)$ in eqn 6.1, which is also what we would expect to see with Newtonian physics. The transition from the quantum world to the Newtonian world occurs when the de Broglie wavelength becomes so small quantum interference can no longer be detected. But at the atomic scale we should anticipate unusual phenomena and processes in materials resulting from quantum physics. One example is quantum diffusion, which is discussed later in this chapter.

The wave-particle duality of electrons is exploited in electron microscopes. Electrons are accelerated and impinge on a specimen. They are scattered by the electrostatic potential inside the specimen. The small de Broglie wavelength of the accelerated electrons enables much finer detail to be resolved than is possible with light in an optical microscope. Accelerated electrons are also able to penetrate much further into materials that are not transparent to visible light. The elastic fields of defects scatter electrons in ways that can be calculated quantum mechanically and compared with experimentally obtained images. The accelerated electrons can also be used to excite electronic transitions in atoms inside the material to establish their chemical identity. Electron microscopy is a standard tool in materials science for imaging as well as for chemical analysis.

Before we leave the double slit experiment let us return briefly to the quantum measurement problem. In Newtonian physics it is normally assumed that when we make a measurement the disturbance of the system is negligible. In quantum physics the system is so small and delicate we must recognise that when we make a measurement we disturb the system significantly. In the case of the double slit experiment this means we cannot introduce any means of determining which slit the photon, electron or C_{60} molecule passes through without at the same time destroying the interference pattern. This intrinsic uncertainty in quantum measurements underpins Heisenberg's uncertainty principle, which is usually stated as follows:

$$\Delta x \Delta p \geq h, \qquad (6.2)$$

where Δx and Δp are the uncertainties in position and momentum of a particle.

In solids the uncertainty relationship leads to the existence of 'zero point motion'. At absolute zero atoms in a solid do not stop vibrating because if they did their position and momentum would be known precisely, which would contravene the uncertainty principle. When quantum mechanics is applied to a harmonic oscillator of frequency ν the minimum energy of the oscillator is $h\nu/2$, which is called the zero point energy. If m is the mass of an atom, and a is the spacing of atoms in the solid then the average momentum p of the atom at absolute zero must be at least h/a and its kinetic

---

[6]Arndt, M, Nairz, O, Vos-Andreae, J, Keller, C, van der Zouw, G and Zeilinger, A, Nature **401**, 680 (1999).

energy at least $p^2/(2m) = h^2/(2ma^2)$. This energy is a lower bound on the zero point energy. In the case of hydrogen interstitials in metals the zero point energy is of order $0.1\,\text{eV}$. To get an idea of how large that is, the metal would have to be heated to about $900°C$ before the thermal energy $k_B T$ becomes comparable to this zero point energy. Although the zero point energy is small in relation to the cohesive energy of most crystals it is often comparable to the difference in cohesive energies of alternative crystal structures. For that reason the zero point energy is usually included in predictions of crystal structure at absolute zero.

6.4   Identical particles, the Pauli exclusion principle and spin

Consider two particles which we label '1' and '2'. Let $A(x_1, x_2)$ be the probability amplitude for particle 1 to be at x_1 and particle 2 to be at x_2. If particle 1 is an electron and particle 2 is a proton they are clearly distinguishable because they have different masses and opposite charges. If we exchange the particles so that the electron is at x_2 and the proton is at x_1 there is no reason why the probability amplitude $A(x_2, x_1)$ should be related to $A(x_1, x_2)$. But suppose the two particles are indistinguishable such as two electrons[7]. Does their indistinguishability mean $A(x_1, x_2)$ is the same as the probability amplitude $A(x_2, x_1)$ for when the positions of the two electrons are exchanged? If we performed an experiment to answer this question we would not measure the probability amplitudes but the probabilities $|A(x_1, x_2)|^2$ and $|A(x_2, x_1)|^2$. Since the electrons are indistinguishable the probabilities would have to be the same. However, that does not necessarily imply that $A(x_2, x_1) = A(x_1, x_2)$ because it could also imply $A(x_2, x_1) = e^{i\phi} A(x_1, x_2)$, where $e^{i\phi}$ is an arbitrary phase factor. However, if we exchange them a second time we must recover the original probability amplitude. It follows that ϕ must be an integer multiple of π, so that the phase factor $e^{i\phi}$ is ± 1. Thus, for indistinguishable particles the probability amplitude has to be either symmetric $A(x_2, x_1) = A(x_1, x_2)$ or antisymmetric $A(x_2, x_1) = -A(x_1, x_2)$ under an exchange of the positions of the particles. Symmetric and antisymmetric probability amplitudes that satisfy these relationships may be constructed as follows[8]:

$$A^{(S)}(x_1, x_2) = \psi(x_1, x_2) + \psi(x_2, x_1) \tag{6.3}$$

$$A^{(A)}(x_1, x_2) = \psi(x_1, x_2) - \psi(x_2, x_1), \tag{6.4}$$

where $\psi(x_1, x_2)$ is the probability amplitude for finding particle 1 at x_1 and particle 2 at x_2, and $\psi(x_2, x_1)$ is the probability amplitude for finding particle 1 at x_2 and particle 2 at x_1. Identical particles with probability amplitudes that are symmetric under an exchange of particle positions are called bosons. Photons are examples of

---

[7]When we say two particles are indistinguishable not only are we saying they have the same physical attributes such as mass and charge, we are also saying we cannot follow their positions and velocities through time owing to the uncertainty principle. Particles obeying Newtonian physics are always distinguishable because their trajectories can be followed. Therefore, for particles to be indistinguishable they have to obey quantum physics not Newtonian physics.

[8]For simplicity the probability amplitudes $A^{(S)}(x_1, x_2)$ and $A^{(A)}(x_1, x_2)$ are not normalized. To normalize them the right hand sides of eqns 6.3 and 6.4 are divided by $\sqrt{2}$, assuming $\psi(x_1, x_2)$ and $\psi(x_2, x_1)$ are normalized.

bosons. Identical particles with probability amplitudes that are antisymmetric under an exchange of particle positions are called fermions. Examples of fermions include electrons, protons and neutrons.

All indistinguishable particles are either fermions or bosons. This classification is based entirely on the symmetry properties of indistinguishable particles following an exchange of their positions. It is yet another example of the fundamental role of symmetry discussed in Chapter 5.

Indistinguishable particles may have an internal degree of freedom called 'spin'. Electrons have one of two possible spins, $+\frac{1}{2}$ and $-\frac{1}{2}$ which are often called spin-up and spin-down. Pauli discovered that fermions have values of spin that are half odd integer: $\pm\frac{1}{2}, \pm\frac{3}{2}, \pm\frac{5}{2} \ldots$ Bosons have whole integer values of spin: $0, 1, 2, \ldots$

The probability amplitude for two electrons involves their spins as well as their positions. However if electrons 1 and 2 have the same spin the only way their probability amplitude can be antisymmetric is for the spatial component to satisfy eqn 6.4. In that case the probability amplitude is zero if $x_1 = x_2$. This means it is impossible for two electrons to occupy the same position if they have the same spin. The antisymmetry of the spatial component of their probability amplitude introduces an effective interaction between the electrons which keeps them apart, in addition to their electrostatic repulsion. This interaction is called the exchange interaction. It is an example of the Pauli exclusion principle, which states that only one electron can occupy a quantum state that includes the specification of spin[9]. Conversely a quantum state excluding the specification of spin can be occupied by at most two electrons, one with spin up and the other with spin down.

6.5   Consequences of the Pauli exclusion principle

The Pauli exclusion principle goes a long way to explaining the chemistry of the elements and the structure of the Periodic Table.

The quantum states of the hydrogen atom can be found analytically. They are characterised by four quantum numbers:

1. $n = 1, 2, 3, 4, \ldots$ is the principal quantum number. In hydrogen the energy of the state is proportional to $1/n^2$ and is independent of the other quantum numbers.
2. $l = 0, 1, 2, \ldots, n - 1$ is the angular momentum quantum number. States with $l = 0, 1, 2, 3$ are called s, p, d, f states respectively.
3. $m = -l, -l+1, \ldots, l-1, l$ is the magnetic quantum number. This quantum number specifies the component of the angular momentum along the z-axis.
4. m_s is the electron spin quantum number.

Each state characterised by n, l, m is called an orbital and may be occupied by two electrons, one with spin up and one with spin down. The nth 'shell' means the set of orbitals sharing the same principal quantum number n. In the nth shell there are n^2 orbitals, accommodating up to $2n^2$ electrons.

---

[9] A quantum state is characterised by a set of quantum numbers which describe values of conserved quantities in the dynamics of an electron, such as its angular momentum in an atom or the linear momentum of a free electron.

Only in hydrogen are the energies of all the orbitals in a given shell the same. In other atoms the electron-electron repulsion favours smaller values of the angular momentum l. As the angular momentum increases the electron spends more time further from the nucleus where its attractive interaction with the nucleus is screened more effectively by other electrons closer to the nucleus. In general, in a given shell the s-state has the lowest energy, then the p-states and then the d-states etc.

Let Z denote the atomic number. Consider the ground-state configuration of electrons in atoms with increasing Z starting from $Z = 1$. The single electron in hydrogen is in a 1s-state. In helium the two electrons can be accommodated in the 1s-state, one with spin up and the other spin down. The first shell is then filled, and helium is an inert gas. The next element ($Z = 3$) is lithium. Its third electron has to go into the second shell, and the lowest energy empty state is the 2s-state. The fourth electron in beryllium completes the filling of the 2s-state. The fifth electron in boron has to go into a 2p-state. The 2p-states are gradually filled in the next elements carbon, nitrogen, oxygen, fluorine and neon. In neon, another inert gas, the second shell is full. From sodium to argon ($Z = 11$ to 18) the 3s and 3p states are gradually filled. The pattern is temporarily broken at $Z = 19$ and 20, which are potassium and calcium, where the 4s-state is occupied before the 3d-states of the first transition metal series are occupied. This is because the 3d-states are spatially very compact so that there is a significant electrostatic repulsion between electrons occupying them and other third shell electrons. The energy of the 4s-state then drops below that of the 3d-states.

The chemistry of the elements is determined by the electrons in the outermost electron shell. The $+1$ valency of the alkali metals is a consequence of the screening of the single outermost electron from the nuclear potential by electrons in lower lying shells. The outermost electron is thus more easily removed by ionisation. The -1 valency of the halogens arises from the single vacancy in the outermost shell to receive an electron and complete the shell. In the inert gases the outermost electrons are in complete shells and their electrons are equally screened from the nucleus by each other and by other electrons in the atom. There is also no unfilled state in the outer shell of electrons in an inert gas. Electrons in the inner filled shells are called 'core' electrons and electrons in the outer partially filled shell are called 'valence' electrons or 'conduction' electrons in a metal. Core electrons do not participate in chemical bonding because their energies are too low, and because their shells are filled – the exclusion principle again. However, the distinction between core and valence electrons cannot always be maintained, for example in the first row of the transition metal series where the 4s-state is occupied before the 3d-states are filled.

When atoms are combined to make molecules the probability amplitude (also known as the wave function) becomes a sum of atomic states with unknown coefficients. The coefficients are found by solving the Schrödinger equation, which also yields the energies of the molecular states. But we can give a flavour of what happens. The core states are too low in energy to be affected by the presence of other atoms and they are not involved in chemical bonding. Molecular states are formed from linear combinations of the atomic valence states. Electrons in these molecular states are itinerant – they are no longer confined to individual atoms as they become delocalised within the molecule. The Pauli exclusion principle still applies: each molecular state may be occupied by at most two electrons, with opposite spins. The quantum numbers characterising the molecular states are no longer the atomic n, l and m numbers.

Macroscopic solids are giant molecules. Imagine the creation of a solid as the coming together of atoms from infinity. The atomic core states remain highly localised and occupied by core electrons in each atom. The valence states become delocalised molecular states in which all atoms in the solid can participate. The energies of the molecular states are in certain ranges called bands. The molecular states are occupied by electrons strictly according the Pauli exclusion principle. Each molecular state is occupied by no more than two electrons of opposite spins, starting from the state with the lowest energy and occupying molecular states of increasing energy until all the valence electrons in the solid are accounted for. Once all the valence electrons in the solid have been assigned to molecular states the remaining molecular states are left unoccupied.

What we have just outlined is called the band theory of solids. It explains why some solids are metallic conductors, some are insulators and some are semiconductors. The energies of the bands of molecular states may overlap or be separated by energy gaps, called band gaps, where there are no molecular states. Full bands are unable to conduct electricity because there are no unoccupied states available to accommodate electrons excited by an applied voltage. Metals have partially filled bands where unoccupied states are easily accessed by exciting electrons in occupied states with the highest energies. Semiconductors have full bands but the energy gap between filled and empty bands is sufficiently small that electrons can be excited thermally from a full band into an empty band.

The existence of bands of electronic states is sometimes connected conceptually to the existence of periodicity in crystals, as if periodicity is a requirement for the existence of bands. If that were true glass would not be transparent. Glass in windows is a non-crystalline insulator. The energy gap between occupied and unoccupied molecular states is sufficiently wide to allow photons at visible frequencies to propagate through the material without encountering electron states they may excite. Amorphous silicon is a semiconductor used in solar cells. It has bands and band gaps, but no crystalline periodicity. The only reason for invoking periodicity in band theory is that it simplifies the calculation of electron energy bands. But it is not a requirement for their existence.

In Section 8.3.1 we will see that sometimes band theory predicts materials are metals when they are in fact insulators. This failure arises from the breakdown of the assumption in band theory that there is no energy cost associated with transferring an electron from one atom to another. The prediction of metallic conductivity in a partially filled band can also fail when there is very strong disorder in the material. Electrons can then become trapped in localised, low energy states and unable to participate in conduction.

In some metals, such as the alkali metals and aluminium the valence electron states lose their atomic character so much they may be approximated as free electron states confined to a box defined by the dimensions of the metal sample. This is the free electron approximation. It is assumed the valence electrons in each atom detach themselves from the atom leaving behind a positive ion. The valence electrons become a sea of electrons free to wander throughout the metal. Despite the strength of the electrostatic interaction it is assumed the valence electrons do not interact with each other or with the positive ions they leave behind. The positive ions are effectively spread out into a uniform positive charge density which neutralises the valence electrons. The electron states of the metal then become standing waves inside the box. In accord with the Pauli exclusion principle each standing wave state is occupied by two electrons, starting from the lowest energy state, and going up in energy until all the valence electrons have been allocated to standing wave states.

The energy of the highest occupied standing wave state is called the Fermi energy and the de Broglie wavelength of electrons in this state is called the Fermi wavelength. The Fermi energy is the kinetic energy of an electron in this highest occupied state. It is a direct result of the Pauli exclusion principle, which forbids the occupation of any state by more than two electrons, that the Fermi energy is of order 10 eV above the bottom of the band. This equates to an electron velocity of around 1% of the speed of light. In other words, even at absolute zero there are electrons in a metal travelling at around a million metres per second.

The Fermi wavelength is comparable to the spacing between atoms in the metal. The Fermi wavelength defines the smallest distance over which the valence electronic charge density in the metal can vary. Suppose we introduce a positive point charge Q into a metal. In a vacuum the electrostatic potential of the point charge decays with the inverse of the distance r from the charge. But in a metal its potential is screened by the free electrons which are attracted to it. They surround the point charge and neutralise it. The size of the neutralising electron cloud surrounding the point charge is comparable to the Fermi wavelength. The electrostatic potential of the screened point charge decays much more rapidly with r. When it is viewed from a distance of a few Fermi wavelengths it appears to be almost neutral and its electrostatic potential is almost zero. The impurity sets up small oscillations in the valence electron density, the amplitude of which decays with distance from the impurity, rather like dropping a stone into a pond.

The free electrons in a metal are extremely effective at screening any charges introduced into it. They are also very effective at screening the electrostatic potential of each other. Each valence electron in the metal is surrounded by a hole in the electronic charge density, the size of which is again comparable to the Fermi wavelength. A depletion in the negative electronic charge density is equivalent electrostatically to a positive charge density. The valence electron is inseparable from the surrounding hole in the valence electronic charge density and together they form a 'quasiparticle', which is neutral. This is the justification for the valence electrons of the metal not interacting with each other electrostatically in the free electron model. The hole surrounding each valence electron forms for two reasons. The first is the exchange interaction ensuring that electrons with the same spin keep out of each other's way. The second is the electrostatic repulsion between electrons. Their combined effect is to ensure that the hole in the surrounding valence electron charge density is equivalent to one missing electron.

In Chapter 1 we met the concept of extensive thermodynamic variables such as the internal energy. Given two macroscopic samples of the same material, the first containing twice as many atoms as the second, the internal energy of the first sample is twice that of the second. This seems obvious and we take it for granted. But why should it be so? Why is there a *linear* dependence on the number of atoms in the sample? It was not until 1967 that the linear dependence was explained theoretically[10]. The same paper showed that the linear dependence was critically dependent on the Pauli exclusion principle. If electrons were bosons instead of fermions the energy of a sample containing N atoms would vary *non-linearly* as $N^{7/5}$. If electrons were not fermions obeying the Pauli exclusion principle matter as we know it would not be stable.

In this section we have seen that the consequences of the Pauli exclusion principle are far-reaching and profound for the properties of all matter. The origin of the principle is the requirement that the probability amplitude for two electrons is antisymmetric when their positions are exchanged. Ultimately, all this follows from the *symmetry* associated with the indistinguishability of identical fermion particles.

6.6   Tunnelling

In Newtonian physics if a particle meets a potential barrier it must have sufficient energy to go over the barrier or it will remain forever on one side. In quantum physics the particle has a finite probability of passing through the potential barrier rather than going over the barrier. This is called quantum tunnelling.

In Section 8.3.1 we will discuss the energy barrier to transferring a valence electron from one atom to another when the atoms are far apart. Unless electrons are able to make such transfers there is no possibility of forming metallic, covalent or ionic bonds. As atoms are brought closer together the energy barriers become smaller and narrower, but they remain insurmountable in Newtonian physics. Eventually electrons are able to tunnel through the barrier. Chemical bonding is a result of quantum tunnelling.

One of the most direct realisations of quantum tunnelling is the scanning tunnelling microscope (STM). An atomically sharp metallic tip is brought within a few Angstroms of the surface of an electrically conducting sample. A small voltage is applied to the tip relative to the sample. Electrons tunnel between the tip and the sample through the potential barrier that separates them. The tip is scanned over the surface. There is a feedback mechanism to maintain a constant tunnel current as the tip moves above the surface by moving the tip towards or away from the surface. The movement of the tip to maintain a constant tunnelling current, as the tip is scanned over the surface, is translated into an image of the surface. The spatial resolution of the STM is such that individual atoms on the surface are readily resolved. The exponential sensitivity (see eqn 6.5 below) of the tunnelling current to movements of the tip towards or away from the surface enables measurements of changes of height at a surface to be made at the picometre (10^{-12} m) level. An example is shown in Fig. 6.2.

6.7   Thermal properties of solids

Although quantum physics was developed for subatomic particles this chapter has shown they affect the properties of materials at the macroscopic level. Macroscopic

---

[10]Dyson, F J and Lenard, A, J. Math. Phys. **8**, 423 (1967).

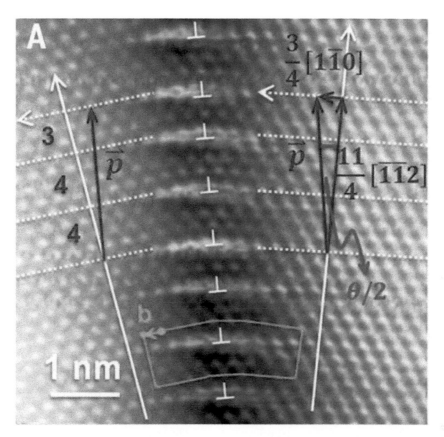

Fig. 6.2 STM image of a (111) surface of a copper thin film with a grain boundary emerging at the surface from top to bottom of the image. Note the 1 nm scale marker: each white circle is an atom. The measured misorientation θ of the grain boundary about the [111] surface normal is 16.2°. The grain boundary contains edge dislocations with Burgers vectors $\frac{1}{2}[1\bar{1}0]$ located at the inverted 'T' symbols. The Burgers circuit in green confirms the Burgers vector of each dislocation. Each $\mathbf{p} = \frac{1}{2}[\bar{4},\bar{7},11]$ period of the boundary contains three edge dislocations. The grain boundary also tilts the (111) surfaces on either side of it by about 3°. As a result the grain boundary is in a shallow valley on the surface of the copper. From Zhang, X, Han, J, Plombon, J J, Sutton, A P, Srolovitz, D J and Boland J J, Science **357**, 397 (2017).. Reprinted with permission from AAAS.

thermal properties of materials, such as the specific heat, are also affected by quantum physics.

In Chapter 1 we saw that heat is the energy of atoms vibrating about their average positions in a solid. The hotter the material, the larger the amplitude of vibration. The molar heat capacity, or 'specific heat', measures the amount of heat required to raise the temperature of one mole of a substance by one kelvin. Newtonian physics predicts the molar specific heat of a solid element is a constant equal to $3R$ at all temperatures up to the melting point, where $R = 8.314\,\mathrm{J\,mol^{-1}K^{-1}}$ is the gas constant. This is known as the Dulong-Petit law, and it is universally violated. As the temperature approaches absolute zero the specific heats of all solids tend to zero. Einstein explained this observation by suggesting that the atoms of a solid may be regarded as independent harmonic oscillators obeying the laws of quantum mechanics. In the Einstein model each atom in a solid sits in a parabolic potential well where it vibrates independently of its neighbours. This is an approximation because atoms are coupled to each other through interatomic forces, so they do not vibrate independently. Nevertheless, the Einstein model captures the essential physics of the failure of the Dulong-Petit law.

It is possible to solve the Schrödinger equation for a harmonic oscillator. If ν is the frequency of vibration of the oscillator its energy is quantised as follows:

$$E_n = \left(n + \frac{1}{2}\right)h\nu,$$

where $n = 0, 1, 2, 3, \ldots$ Notice the lowest energy state is $E_0 = \frac{1}{2}h\nu$, which is the zero point energy referred to previously. A typical frequency of vibration of an atom in a solid is around $10^{12} - 10^{13}\,\mathrm{s^{-1}}$. For this range of frequencies $h\nu$ is between 0.004 and 0.04 eV. If the thermal energy available is significantly less than $h\nu$ it is unlikely the atomic oscillators will be excited from E_0 to E_1. In that case the atomic oscillators are unable to absorb thermal energy and the specific heat decreases as the temperature decreases, reaching zero at absolute zero. At low temperatures, the discrete energy levels of the harmonic oscillators are separated by too much energy for excitations to higher energy states to occur.

The Debye model of the specific heat is an improvement on the Einstein model in that it takes into account the coupling between atoms in the solid. The coupling leads to a distribution of frequencies of vibrations which depend on their wavelengths. The Debye model also includes the quantisation of the energy levels of the various collective modes of vibration. These quantised collective modes of vibration are called phonons. At low temperatures in the Debye model the specific heat arising from the excitation of atomic vibrations is predicted to vary as T^3, in agreement with experiment. At high temperatures the specific heat in the Debye and Einstein models tends to the value of $3R$ predicted by the Dulong-Petit law. The temperature T_D at which the dominant physics of the specific heat changes from quantum to classical is called the Debye temperature T_D. It is defined by $h\nu_{max} = k_B T_D$, where ν_{max} is the maximum frequency of the modes of vibration of atoms in the crystal, and k_B is the Boltzmann constant. Atoms with small masses bound by strong interatomic bonds have high vibration frequencies. For this reason the Debye temperature in diamond is more than $2,000\,\mathrm{K}$. We see that quantum mechanics directly affects the *macroscopic* thermal properties of a solid.

In metals the valence electrons can also contribute to the specific heat. Electrons near the Fermi energy can be excited to higher energy unoccupied states only slightly above the Fermi energy. But the majority of valence electrons in the metal are unable to be excited to higher energy states because there are no unoccupied states they can access. The number of electrons that can be excited is thus limited by the Pauli exclusion principle to those just below the Fermi energy that can be excited by the thermal energy available. For this reason the electronic contribution to the specific heat in a metal varies linearly with temperature. At most temperatures the specific heat in a metal is dominated by the vibrational contribution. But at very low temperatures the electronic contribution dominates owing to its linear dependence on temperature, while the vibrational contribution has a cubic dependence on temperature.

6.8 Quantum diffusion

We have seen that to understand the behaviour of electrons in solids we have to treat them quantum mechanically. This raises the interesting question of whether and when we have to treat the dynamics of the atomic nuclei quantum mechanically. In the previous section we have seen that quantisation of atomic vibrations has a strong influence on the thermal properties of solids at temperatures below the Debye temperature. Atoms in a solid also display zero point motion, which is a feature of quantum physics. Zero point motion also contributes to the relative stabilities of alternative crystal structures.

The treatment of diffusion outlined in Chapter 3 is not based on quantum physics. A thermal fluctuation provides an atom adjacent to a vacant site with sufficient energy to *overcome* a potential barrier and it jumps into the vacancy. Similarly an interstitial receives sufficient energy from a thermal fluctuation to *overcome* a potential barrier into a neighbouring interstitial site. Could the diffusing atom tunnel through the barrier rather than go over it?

This is a question of considerable technological significance, especially for the burgeoning 'hydrogen economy'. Hydrogen is a notorious embrittling element in many metals and alloys, including steels. It can be extremely difficult to remove hydrogen atoms from a material and to prevent their ingress, for example from the dissociation of water. They are attracted to regions of relatively high tensile stress where they may nucleate cracks. Consequently their mobility is a matter of concern. In practice alloy designers try to build into the alloy traps for hydrogen atoms to immobilise them. But a 'trap' is just a deeper potential well for the hydrogen atom, and again the question arises as to whether it can tunnel through the potential barrier surrounding the trap.

Consider a free particle of mass M and energy E travelling along the x-axis where it meets a potential barrier $V(x)$ between $x = 0$ and $x = w$. The barrier is given by $V(x) = V_0 \sin^2(\pi x/w)$, where $V_0 \gg E$. In Newtonian physics the particle would be reflected by the barrier. But in quantum physics the probability of the particle tunnelling through the barrier is given approximately by:

$$P \approx e^{-8\sqrt{2MV_0}\,w/h}. \tag{6.5}$$

For a hydrogen atom, with $V_0 = 0.5\,\text{eV}$ and $w = 1\,\text{Å}$ we find $P \approx 3 \times 10^{-9}$. If the barrier width is doubled, or the barrier height is increased to $2\,\text{eV}$, or the hydrogen is replaced by helium, the probability of tunnelling is reduced approximately to 9×10^{-18}. How do we interpret these numbers?

If a hydrogen atom is vibrating within an interstice of diameter approximately $1\,\text{Å}$ the uncertainty principle requires it has a minimum momentum of $\approx 7 \times 10^{-24}\,\text{kg m s}^{-1}$.

Therefore the hydrogen atom has an average minimum kinetic energy of about 0.08 eV. If we equate this to the zero point energy, $h\nu/2$, of the hydrogen atom, its frequency ν of vibration is at least $4 \times 10^{13}\,\mathrm{s}^{-1}$, which looks reasonable. Thus, the hydrogen atom makes at least 4×10^{13} attempts per second to tunnel through the barrier. If the probability of tunnelling is at least $P \approx 3 \times 10^{-9}$ then more than 10^5 tunnelling events will take place per second. These tunnelling events trace a random walk between interstitial sites in three dimensions. The distance travelled in one second is at least $\sqrt{10^5}\,\text{Å} \approx 300\,\text{Å}$ (assuming the distance between interstitial sites is $1\,\text{Å}$). Equating this to $\sqrt{Dt}$ with $t = 1\,s$, we obtain a lower bound for the diffusivity $D \approx 10^{-15}\,\mathrm{m^2\,s^{-1}}$.

The above estimate is very over simplified. By invoking an attempt frequency it mixes classical and quantum physics. A more consistent treatment in quantum theory would be to solve the Schrödinger equation for a hydrogen atom in a periodic potential representing the periodic distribution of interstitial sites in a crystal. It also overlooks an important requirement for tunnelling to occur, namely the conservation of energy: the energies of the system before and after tunnelling must be the same. When the hydrogen atom is in an interstitial site the surrounding metal atoms relax to slightly different positions as a result of the forces the hydrogen atom exerts upon them. These atomic displacements reduce the energy of the system before tunnelling occurs. Therefore, for tunnelling to occur it is likely that displacements of host atoms have to take place around the initial and final interstitial positions to ensure conservation of total energy. These atomic displacements may be thermally assisted.

Nevertheless, eqn 6.5 suggests that hydrogen diffusion by quantum tunnelling is feasible. It also suggests that tunnelling is less likely with greater atomic mass.

6.9   Closing remarks

The introduction of quantum mechanics at the beginning of the twentieth century marks one of the greatest revolutions in the history of science. It is sometimes said that its impact on materials science is confined to the atomic scale – to understanding bonding and electronic, magnetic and optical properties of materials. But it permeates the whole of materials science, contributing to the free energies of phases as a function of temperature and composition, the free energies of formation and migration of defects, and the drag forces they experience. It is hard to think of any area of materials science where quantum mechanics does not have at least an indirect influence.

Further reading

Feynman, R P, Leighton, R B and Sands, M, *Feynman lectures on physics Volume 3: Quantum Mechanics*, Addison Wesley Publishing Company (1965).

Susskind, Leonard and Friedman, Art

Quantum Mechanics: The theoretical minimum, Penguin Books (2014).

Sutton, Adrian P, *Electronic structure of materials*, Oxford University Press (1996).

7
Small is different

Up to now, we have been content to dig in the ground to find minerals. We heat them and we do things on a large scale with them, and we hope to get a pure substance with just so much impurity, and so on. But we must always accept some atomic arrangement that nature gives us. We haven't got anything, say, with a "checkerboard" arrangement, with the impurity atoms exactly arranged 1,000 angstroms apart, or in some other particular pattern.

What could we do with layered structures with just the right layers? What would the properties of materials be if we could really arrange the atoms the way we want them? They would be very interesting to investigate theoretically. I can't see exactly what would happen, but I can hardly doubt that when we have some control of the arrangement of things on a small scale we will get an enormously greater range of possible properties that substances can have, and of different things that we can do.

Feynman, R P, *There's plenty of room at the bottom*, February 1960 issue of *Engineering and Science*, published by California Institute of Technology. Transcript of a talk Feynman gave at the American Physical Society annual meeting on December 29th 1959.

7.1 Concept

Materials with nanoscale dimensions behave differently from their macroscopic counterparts. Their properties are more obviously dominated by quantum physics. The development of techniques to create and manipulate materials at the nanoscale has underpinned the current age of information.

7.2 Introduction

Feynman's visionary talk in 1959 was largely forgotten until 1981 when the scanning tunnelling microscope was invented[1]. This was the breakthrough that marked the beginning of nanoscience, which led eventually to our ability to share, store and retrieve inconceivable amounts of information in portable devices. Feynman's dream of creating

---

[1]Binnig, G, Rohrer, H, Gerber, Ch. and Weibel E, Phys. Rev. Lett. **49**, 57 (1982).

materials at the nanoscale in pre-determined patterns has now been realised, including his checkerboard[2] and multilayer[3] examples.

The nanoscale is usually defined as the range of length scales from 10^{-9} m (1 nm) to 10^{-7} m (100 nm or $0.1\,\mu$m). When any external dimension of a material is at the nanoscale it is called a nanomaterial. The focus of this chapter is the science of nano-materials, which is part of the field of nanoscience. There are also nanostructured materials. Examples include nanocrystalline materials[4] in which the grain size is at the nanoscale, and other materials containing particles or microstructural features of nanoscale dimensions.

Consider how the properties of a cube of a crystal of pure gold change as we reduce the length L of the cube edge. When $L = 1$ cm the values of its internal energy, entropy and volume are directly proportional to the number of atoms in the cube. Therefore these properties satisfy the definition of an extensive thermodynamic variable. But not quite, because atoms at the surface of the cube have different numbers of neighbouring atoms from those in the interior. Therefore the contributions surface atoms make to these thermodynamic variables are expected to differ from those of atoms in the interior of the cube. A simple calculation shows that when $L = 1$ cm the ratio of the number of surface atoms to interior atoms is of order 10^{-8}. Therefore the contributions of surface atoms to these thermodynamic variables is negligible and we are justified in treating them as extensive variables. When the same argument is applied to other physical properties of the cube, such as its density, electrical conductivity, optical reflectance, elastic constants etc., it is concluded that they are indistinguishable from those of a much larger samples under the same environmental conditions.

As the size of the cube decreases the ratio of the number of surface atoms to the number of interior atoms increases. For example, when $L = 100$ nm the ratio is about 0.6% and when $L = 10$ nm the fraction of atoms at the surface is about 6%. One of the consequences of this trend is that the melting point is reduced because there is a higher fraction of atoms in higher energy sites. Another is that the density of the sample increases. If we assume the energy per unit area, γ, of the surface is a constant the surface energy of the cube is reduced if the total surface area of the cube is smaller – the surface energy puts the cube into compression. The value of the resulting compressive stress inside the cube is of order γ/L. This is a rough estimate because γ is an average taken over a relatively large patch of surface, and it depends on the orientation of the surface normal relative to the crystal axes. It is also likely to change when L decreases to $1 - 10$ nm because a significant number of atoms are at cube edges where two surfaces meet. Setting $\gamma = 1\,\mathrm{J\,m^{-2}}$, which is the order of magnitude of a typical surface energy of a metal, we find the compressive stress is of order 10 MPa when $L = 100$ nm, 100 MPa when $L = 10$ nm and 1 GPa when $L = 1$ nm. As a result of this compressive stress nanoparticles have smaller atomic separations than their bulk counterparts. Some nanoparticles adopt different crystal structures as

---

[2] MacManus-Driscoll, J, Zerrer, P, Wang, H, Yang, H, Yoon, J, Fouchet, A, Yu, R, Blamire, M G and Jia, Q, Nature Mater **7**, 314 (2008).

[3] Esaki, L, IEEE Journal of Quantum Electronics **22**, 1611 (1986).

[4] for an illuminating review see Meyers, M A, Mishra, A and Benson, D J, Prog. Mat. Sci. **51**, 427 (2006).

they decrease in size[5]. This simple example of pure gold shows that as the size of a material decreases to the nanoscale its properties become size-dependent because of the growing influence of surface atoms.

There are other interesting changes in the properties of a sample of gold when its size reaches the nanoscale. One of them is its chemical reactivity. Gold is a 'noble' metal because it does not react with oxygen or become corroded in moist air. However, particles of gold less than 5 nm in diameter supported on selected oxide particles have been found to be excellent catalysts for certain chemical reactions[6]. Furthermore, suspensions of gold nanoparticles in water may be coloured red, orange or blue depending on the shape and size of the particles[7]. These are indications that the electronic properties of gold nanoparticles are not the same as those of bulk samples.

Feynman noted that as the size of a particle approaches the atomic scale its properties become more obviously determined by quantum physics. A good example, mentioned in Section 8.3.1, is the transition from diffusive to ballistic electron transport, with the concomitant quantisation of the conductance, in atomically smooth metallic wires as they become narrower. In this chapter we encounter other properties of nanomaterials that have their roots in quantum physics.

7.3 Quantum dots

Quantum dots are small crystals typically a few nanometres across. An electron in a quantum dot is confined to the small volume of the crystal. Its possible quantum states may be determined approximately by treating it as an electron confined within an infinitely deep three dimensional potential well representing the particle. This is called the particle-in-a-box model, where the box is the potential well. It is a standard exercise for undergraduates to solve the Schrödinger equation for a particle confined to a cube. The possible energies of an electron confined to a cube of side L are as follows:

$$E(n_x, n_y, n_z) = \frac{h^2}{8mL^2} \left(n_x^2 + n_y^2 + n_z^2 \right), \tag{7.1}$$

where $n_x, n_y, n_z = 1, 2, 3, 4, \ldots$, and m is the mass of the electron. The energy levels are quantised because they are determined by the numbers n_x, n_y, n_z, which may assume only integer values. The energy levels are proportional to $1/L^2$. As L decreases the separations between successive quantised energy levels increases. This simple model ignores the variation of the potential within the quantum dot and it ignores the interactions between the electrons, yet it is remarkably successful at capturing what is observed experimentally especially if the mass m is replaced by an effective mass.

---

[5] Gold nanoparticles appear to retain essentially the same crystal structure as bulk gold down to diameters of 3 nm, but they display increasing levels of local structural disorder as their size decreases – see Petkov, R, Peng, Y, Williams, G, Huang, B, Tomalia, D, and Ren, Y, Phys. Rev. B **72**, 195402 (2005).

[6] Haruta, M, Catalysis Today **36**, 153 (1997).

[7] first discovered by Faraday by reducing a solution of gold chloride in water with phosphorus – see Faraday, M, Phil. Trans. R. Soc. **147**, 145 (1857).

For a given volume of the quantum dot the energy levels are sensitive to its shape[8] because its shape determines the boundary conditions that have to be satisfied by the wave function in the Schrödinger equation.

Each available energy level in eqn 7.1 can be occupied by at most two electrons, one with spin up and the other spin down. Thus each quantum state is characterised by four quantum numbers: n_x, n_y, n_z, s, where s labels the spin and is either 'up' or 'down'. Some of the energy levels are degenerate, which means they have the same energy but different sets of the three integers n_x, n_y, n_z. For example, the states with $(n_x, n_y, n_z) = (2, 1, 1), (1, 2, 1), (1, 1, 2)$ are degenerate.

The discreteness of the energy levels in eqn 7.1 indicates that when an electron moves from one quantum state characterised by the quantum numbers $(n_{x1}, n_{y1}, n_{z1}, s)$ to another quantum state characterised by $(n_{x2}, n_{y2}, n_{z2}, s)$ there is a finite amount of energy involved in the transition. For the transition to be possible the Pauli exclusion principle requires that the state $(n_{x2}, n_{y2}, n_{z2}, s)$ is unoccupied before the transition is attempted.

We have noted already that in this model the potential felt by electrons inside the quantum dot is zero. This means that inside the quantum dot the electrons are free in the sense that they are not bound to any particular atom. As discussed in Section 6.5 this model is a reasonable description of free electron metals such as aluminium.

In practice quantum dots are made from materials such as silicon, germanium, zinc oxide, indium phosphide and indium arsenide. In bulk form these materials are semiconductors with a full valence band separated by a gap in energy from an empty conduction band (see Section 6.5). In a cubic centimetre of these materials the numbers of states in the valence and conduction bands are of order 10^{23}. The difference in energy between successive states in each band is therefore negligible and there is effectively a continuum of states as a function of energy within each band.

However, as the crystal size decreases to the nanoscale the states in the valence and conduction bands become discrete, as in eqn 7.1. At the same time the widths of the valence and conduction bands decrease slightly. Consequently, the gap, Δ, between the highest occupied state and the lowest unoccupied state increases slightly, as illustrated in Fig. 7.1.

If an electron is excited from the highest occupied state into the lowest energy unoccupied state it will leave behind a 'hole' – a missing electron. The nanocrystal is electrically neutral. Since the excited electron is negatively charged the hole it leaves behind must be positively charged to maintain charge neutrality. Being oppositely charged the electron and hole attract one another and together they form an 'exciton'. The other charges in the nanocrystal screen (i.e. reduce) the electrostatic attraction between the excited electron and its hole. But they don't eliminate it altogether and the small size of the nanocrystal prevents the electron and hole separating far. The exciton has a negative energy because to separate the electron and hole requires work to be done to overcome their mutual attraction. Therefore, when an electron is excited from the highest occupied state into the lowest unoccupied state the resulting exciton has an energy E_{exciton} slightly less than Δ. It follows that the frequency of light emitted

---

[8]Problem 28 on p247 of my book *Electronic structure of materials* compares the energy levels of cubic and spherical quantum dots with the same volume.

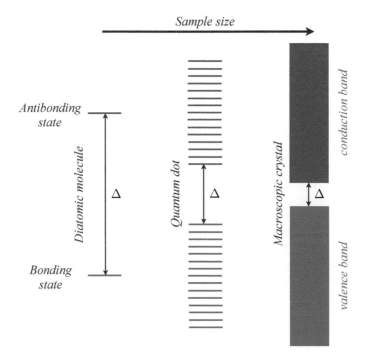

Fig. 7.1 Schematic illustration of the changes in the energy levels of a semiconductor with sample size. Δ is the difference in energy between the highest occupied state and the lowest unoccupied state. For the diatomic molecule Δ is the difference in energy between the bonding and antibonding states. For the macroscopic crystal Δ is the band gap energy and the states are continuously distributed in the valence and conduction bands. For the quantum dot Δ is between these two limits, and its states are separated by finite energies so that they are 'discrete'.

when the electron and hole in the exciton recombine is determined by $h\nu_{out} = E_{\text{exciton}}$. By carefully controlling the size and shape of the nanocrystal the exciton energy can be made to correspond to light of different colours.

Electrons in the quantum dot may be excited by illumination with white light, which contains a spectrum of frequencies. An electron in an occupied state of energy E_1 may be excited into an unoccupied state of energy E_2 if the illuminating light contains a frequency ν_{in} given by $h\nu_{in} = E_2 - E_1$. The energy difference $E_2 - E_1$ is larger than the exciton energy, so that the emitted light is of a lower frequency than ν_{in}. The energy $h\nu_{in} - h\nu_{out}$ is positive and it is either converted into heat by exciting atomic vibrations or it is emitted as infra-red radiation. The process of absorbing light of a higher frequency and emitting light of a lower frequency is called fluorescence.

Quantum dots have found a wide range of applications in nanotechnology[9]. They

[9]e.g. *The many aspects of quantum dots*, Nature Nanotech **5**, 381 (2010).

have also found applications in medicine where they have been used for example to deliver drugs and for imaging purposes[10]. It is clear there have to be limits on the variability of shapes and sizes of the quantum dots, and their chemical composition. One of the most successful ways of controlling the size, shape and composition of quantum dots is to use colloidal chemistry to synthesise them[11].

7.4 Catalysis

Catalysts increase the rate of a chemical reaction without being consumed by the reaction. Sometimes different reaction products result from the same reactants by using different catalysts. The selectivity of a catalyst is the ratio of the amount of desired product to the amount of undesired product, where the amounts are measured in moles. An example is the reaction between carbon monoxide (CO) and hydrogen gas (H_2). If the reaction is carried out with a nickel catalyst the products are methane (CH_4) and water. But if the reaction is carried out with a zinc/copper catalyst the product is methanol (CH_3OH). The shape and size of a nanocrystal affects both its reactivity and selectivity as a catalyst. The shape determines which crystal facets are at the surface. It also determines the arrangement of edges and corners which may be special sites where the reaction occurs.

Metal nanocrystals have been grown with remarkable control over their shapes and sizes by using solution chemistry[12]. This has enabled some careful studies to be made of the relationship between the reactivity of a noble metal catalyst and its shape and size[13]. However, chemistry is highly specific and it is very difficult to draw general conclusions.

7.5 Giant magnetoresistance

7.5.1 The origin of magnetism

Before I describe the giant magnetoresistance effect I will give an outline of where magnetism comes from. Magnetic fields are created by moving charges, as when a current flows along a wire. Consider a very small loop of wire of area δA carrying a current I. The magnitude of the magnetic dipole moment associated with the loop is defined by $I\delta A$. The magnetisation of a permanent magnet is defined as the magnetic dipole moment per unit volume. The magnetic field generated by the small current-carrying loop is equivalent to that of a small permanent magnet of volume δV in which the magnetisation is $I\delta A/\delta V$.

Consider the magnetic dipole moment of a single point charge q of mass m in a circular orbit. The 'current' is the charge q divided by the time taken for the charge to complete one orbit. If r is the radius of the orbit and v is the speed of the charge this time is $2\pi r/v$. Therefore the current is $qv/(2\pi r)$. The area of the orbit is $\delta A = \pi r^2$. Therefore, the magnetic moment of the orbiting charge is $qvr/2$. The magnitude of the

[10]e.g. Parveen, S, Misra, R, Sahoo, S K, Nanomedicine: Nanotechnology, Biology and Medicine **8**,147 (2012).

[11]Kovalenko, M V, Liberato, M, Cabot, A *et al.*, ACS Nano **9**, 1012 (2015).

[12]Xia, Y, Xiong, Y, Lim, B, Skrabalak, S E, Angewandte Chemie **48**, 60 (2009).

[13]Du, Y, Sheng, H, Astruc, D and Zhu, M, Chem. Rev. **120**, 526 (2020).

angular momentum of the charge is $L = mvr$. The magnetic moment of the orbiting charge is thus γL where $\gamma = q/(2m)$ is called the gyromagnetic ratio. This is a simple derivation based on classical physics. The key point is that there is a close connection between the magnetic moment of an orbiting charge and its angular momentum, and that point carries over in quantum theory. That is as far as classical physics can take us because magnetism can be understood only in quantum theory.

In quantum theory the angular momentum is quantised, so that it has only certain discrete values. In addition to the angular momentum arising from its orbital motion around the nucleus an electron also has angular momentum arising from its spin. The existence of electron spin is a result of relativistic quantum theory. Quantum theory shows that the gyromagnetic ratio for the spin of an electron is twice as large as it is for its orbital motion, i.e. it is e/m, where e is the magnitude of the electron charge. Its total angular momentum is the sum of the orbital and spin angular momenta, taking into account their directions as well as their magnitudes, because angular momentum is a vector quantity. The resultant angular momentum of the electrons in an atom is the vector sum of their individual angular momenta. Owing to the linear relationship between the magnetic moment of each electron and its angular momentum the overall magnetic moment of an atom is the vector sum of the magnetic moments of each electron in the atom. Quantum theory also shows that the angular momenta of electrons in filled shells add up to zero. Therefore, only the electrons in unfilled shells contribute to the magnetic moment of an atom and they are the *valence electrons*. For example, the inert gas atoms do not have permanent magnetic moments because they have no unfilled shells of electrons.

In crystals the orbital angular momentum is often 'quenched'. Its contribution to the angular momentum of electrons *in the crystal* is then zero, and the total angular momentum is determined only by the electron spin contribution. This is the case in the transition metals. The reason for the quench is the change of symmetry of the environment of the atom when it is in the crystal as compared to the spherical symmetry of a free atom. In the reduced rotational symmetry of an atomic site in a crystal all the components of the orbital angular momentum may average to zero, even though the total orbital angular momentum may be quite accurately conserved. Since the component of the orbital angular momentum in any particular direction averages to zero its contribution to the total magnetic moment is also zero.

Most free atoms possess a permanent magnetic moment. That is because the exchange interaction between valence electrons in a free atom favours parallel alignment of their spins. This is the basis of Hund's rule for atoms. In that case the Pauli exclusion principle keeps the electrons further apart than they would otherwise be inside the atom, thereby reducing their electrostatic repulsion and hence reducing the total electronic energy of the free atom.

When atoms come together and form bands of electronic states the exchange interaction between electrons on adjacent atoms favours alignment of their spins. It competes with the electronic kinetic energy which favours antiparallel spins. If the exchange energy is sufficiently large the material is ferromagnetic like iron. This occurs only in narrow bands like the 3d band in transition metals. At room temperature

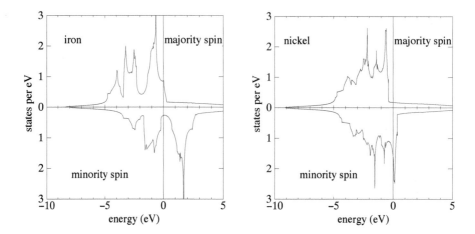

Fig. 7.2 Calculated densities of states for majority and minority electron spins for iron and nickel as a function of the electron energy in eV. The Fermi energy is at 0 eV. The assumed lattice constants for body centred cubic iron and face centred cubic nickel were 2.87 and 3.52 Å respectively. Calculated by Professor Tony Paxton using density functional theory in the local spin density approximation of U von Barth and L Hedin J Phys C: Solid State Phys. **5**, 1629 (1972), as modified by V L Moruzzi, J F Janak and A R Williams, *Calculated electronic properties of metals*, Elsevier (1978).

only the elements iron, cobalt, nickel and gadolinium are ferromagnetic crystals[14]. In these metals there is a spontaneous alignment of the spins of 3d electrons below a critical temperature, called the Curie temperature. Thus, a ferromagnetic material has a permanent magnetisation below the Curie temperature, as in a familiar iron magnet. Thermal agitation reduces the degree of the spin alignment. Above the Curie temperature the permanent magnetisation disappears as the orientations of the electron spins no longer align over more than a few atomic distances. Above the Curie temperature the atoms still have magnetic moments but entropy has taken over and they are no longer aligned over significant distances. The state of the material is then paramagnetic.

The origin of ferromagnetism is a majority of electrons in the metal with one spin and a minority with the opposite spin. The magnetisation of the metal is proportional to the difference, per unit volume, between the numbers of electrons with majority and minority spins. The direction of the magnetisation is not uniquely defined until we apply a magnetic field. At absolute zero the majority spins are parallel to the magnetic field and the minority spins are antiparallel to it. If the direction of the magnetic field is reversed the populations of the majority and minority spins are flipped and the magnetisation reverses direction. The Fermi energy is the same for majority and minority spin electrons. It is the highest energy of occupied states and is the same throughout the metal.

---

[14]In the case of gadolinium the Curie temperature is 20°C, so its inclusion in this list depends on the definition of 'room temperature'.

In Fig. 7.2 I show the densities of states for majority and minority spin valence[15] electrons on each atom in iron and nickel. The density of states tells us how the number of electronic states with a particular spin varies with the energy of the states. A larger density of states at a particular energy means there are more states at that energy. In a nonmagnetic material the densities of states are identical for states with either spin. The zero of energy is arbitrary and in Fig. 7.2 it is equated to the Fermi energy. States with energies below the Fermi energy are occupied, those above are unoccupied[16]. The total number per atom of valence electrons with a particular spin is the area under the density of states curve up to the Fermi energy. The calculated numbers per atom of majority and minority spin valence electrons are respectively 5.14 and 2.87 for iron and 5.31 and 4.69 for nickel. The differences between the numbers of majority and minority spins are 2.27 and 0.62 for iron and nickel respectively. These calculated values compare well with the experimentally observed values of 2.2 and 0.6. Notice that the densities of states at the Fermi energy are unequal for the majority and minority carriers in each metal, and that there are more states above the Fermi energy for the minority spins than the majority spins.

In a macroscopic ferromagnet the magnetisation is not the same everywhere because there are domains where it changes orientation. The existence of domains reduces the energy stored in the magnetic field created by the magnet. But the boundaries between adjacent domains have an energy cost and the trade-off between these two energies determines the average size of the domains. As mentioned before, a crucial property of ferromagnets is that the direction of the magnetisation can be altered by changing the direction of an external magnetic field. In a macroscopic ferromagnet this is achieved by movement of the domain boundaries such that domains in which the orientation of the magnetisation is closer to that of the magnetic field grow at the expense of others. The existence of domains, with their associated energies and mobilities, is a further example of the multiscale nature of materials science.

7.5.2 Magnetoresistance in ferromagnetic metals

Consider a ferromagnetic state of a metal in which there is just one domain. In 1936 Nevill Mott gave a simple explanation of the observed change of the temperature dependence of the resistivity of a ferromagnetic metal at the Curie temperature[17]. Current-carrying electrons are injected into the metal from a battery. Half of them may be considered to be parallel to the magnetisation of the metal and the other half antiparallel. All of them encounter scattering centres in the metal such as vibrating atoms and structural defects. The spin of a current-carrying electron is unlikely to change during the scattering process. Therefore, to a good approximation the spin parallel and spin antiparallel current-carrying electrons may be treated as two independent streams. For an electron to be scattered the Pauli exclusion principle requires

---

[15] Recall that 'valence' electrons are those in partially occupied shells of the atom. The numbers of electrons with each spin are equal in full shells. Therefore, we need to consider only the valence electrons to understand where ferromagnetism comes from.

[16] Strictly speaking this is true only at absolute zero, but it is a very good approximation at room temperature.

[17] Mott, N F, Proc. R. Soc. Lond. A **153**, 699 (1936).

that there is an unoccupied electronic state in the metal to receive it after it has been scattered. Since the spin of the electron is unlikely to change after it is scattered it is likely the receiving unoccupied state must have the same spin. If there are more unoccupied states at the Fermi energy with spin antiparallel then current-carrying electrons with spin antiparallel are scattered more than those with spin parallel, and *vice versa*. It follows that the resistances for spin parallel and antiparallel current-carrying electrons are not the same in a ferromagnet.

7.5.3   The giant magnetoresistance effect

Computer hard drives store and retrieve information encoded in binary format as 'bits', each of which is one or zero. Each bit of information occupies a small area on the disc which is magnetised in two possible ways, one of which corresponds to 'one' and the other to 'zero'. The information is transferred to the disc by a 'write head' in which current pulses magnetise areas of the disc as it spins under the head. Information is retrieved from the disc by changes in the current passing through a 'read head'. As magnetised regions of the spinning disc pass under the read head they alter the magnetisation in the head. The associated changes in magnetoresistance alter the current flowing through the read head. Changes in the current are interpreted as ones and zeros and thus the binary information is retrieved.

To store more information on a hard disc of given size the area allocated to each bit of information has to shrink. But then the magnetic field associated with each magnetised area decreases. It becomes more difficult to detect these weaker magnetic fields unless the sensitivity of the read head is increased. That is what the giant magnetoresistance (GMR) effect enabled.

Discovered in 1988 independently by Albert Fert and Peter Grünberg, GMR revolutionised the technology of magnetic information storage on hard discs. Fert and Grünberg were awarded the Nobel Prize in Physics in 2007 for their discovery. Their work also created fresh impetus in the field of spintronics, where electron spin is used to carry and store information in nanoscale devices.

In its simplest configuration the GMR effect consists of a thin non-magnetic (e.g. chromium) layer sandwiched between two ferromagnetic (e.g. iron) layers, as shown in Fig. 7.3. The thicknesses of these layers are just a few nanometres. They are grown atomic layer by atomic layer using molecular beam epitaxy on a single crystal substrate of gallium arsenide. Current is passed between the terminals on the top surface and so travels through the layers, as shown schematically in Fig. 7.3a.

In the absence of an applied magnetic field ($H = 0$) the magnetisations in the ferromagnetic layers are opposite to each other. This is a consequence of a coupling between the magnetic layers through the intermediate nonmagnetic layer. The coupling is quantum mechanical in origin. Its sign varies with the thickness of the nonmagnetic layer so that the magnetisations in the ferromagnetic layers can also be the same in the absence of an applied magnetic field. However, the GMR effect exists only when the coupling leads to antiparallel magnetisations in the ferromagnetic layers.

Fig. 7.3b shows schematically the change of resistance of the trilayer as a function of the applied magnetic field. In the absence of an applied field the resistance is R_{AP}. When the applied magnetic field aligns the magnetisations the resistance falls to R_P.

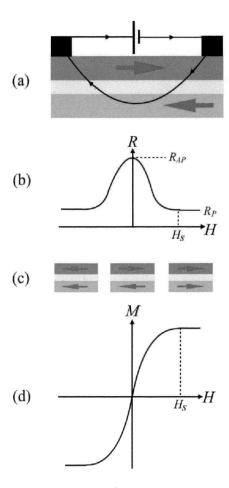

Fig. 7.3 Schematic illustration of the GMR effect. (a) Current-carrying electrons (black arrows) pass through the trilayer comprising two ferromagnetic layers (blue and green) separated by a nonmagnetic layer (yellow). The red arrows show the magnetisation in each ferromagnetic layer. In the absence of an applied field the magnetisations are antiparallel. (b) The resistance R of the trilayer as a function of the applied magnetic field H. R_{AP} (R_P) is the resistance when the magnetisations are antiparallel (parallel). H_s is the saturation magnetic field. (c) The magnetisation configurations at saturation (left and right), and with no applied field (middle). (d) The magnetisation M of the trilayer as a function of the applied magnetic field.

Fig. 7.3c shows the magnetisations in the trilayer at zero and saturation (H_s) applied fields. The resultant magnetisation of the trilayer is shown in Fig. 7.3d. The ratio of the change in the resistance $R_{AP}-R_P$ to R_P is how the effect is usually characterised. Since the magnitude of the effect was so much larger than previous field-induced variations of magnetoresistance it was called 'giant'.

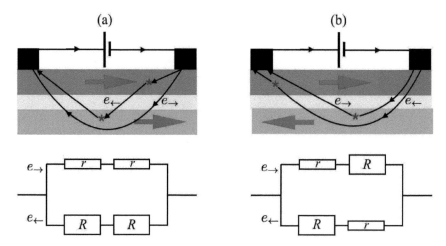

Fig. 7.4 Explanation of the GMR effect. The purple stars signify scattering events. (a) Magnetisations are parallel. Current carrying electrons with spins parallel to the magnetisation encounter small resistances r in the ferromagnetic layers. But those antiparallel to the magnetisation meet larger resistances R in each ferromagnetic layer. (b) Magnetisations are antiparallel. Current carrying electrons of either spin encounter a resistance $R + r$. The resistance for the parallel magnetisation configuration, $2rR/(R+r)$, is always less than that for the antiparallel configuration, $(R+r)/2$, provided $R \neq r$.

The explanation[18] of the GMR effect is illustrated in Fig. 7.4. Invoking Mott's argument we assume the scattering probability for electrons with spin antiparallel to the magnetisation is greater than for electrons with spin parallel to it. We shall find that if the relative scattering probabilities are the other way round the argument still stands. The argument hinges only on the scattering probabilities being different for the two spin states.

When the magnetisations in the ferromagnetic layers are parallel, current-carrying electrons with majority spins pass through the trilayer meeting only relatively small resistances r in each ferromagnetic layer, see Fig. 7.4a. However, electrons with minority spins encounter larger resistances R in each ferromagnetic layer. But when the magnetisations are antiparallel, electrons with either spin meet resistance r in one ferromagnetic layer and resistance R in the other, see Fig. 7.4b. The equivalent resistor configurations for the two streams of current-carrying electrons in the parallel and antiparallel magnetisation configurations are also shown. The overall resistance in the antiparallel configuration is $R_{AP} = (R+r)/2$, and in the parallel configuration it is $R_P = 2Rr/(R+r)$. As long as R and r differ, R_P is always less than R_{AP}. In conclusion, there is always a magnetoresistance effect in the trilayer as long as current-carrying electrons encounter a resistance that depends on their spin state.

Translating this fundamental discovery into commercial products took less than ten years. It led to a dramatic increase in the density of information stored on hard

---

[18] based on Tsymbal, E Y and Pettifor, D G, Solid State Physics **56**, 113 (2001).

discs. GMR read heads have now been replaced by tunnel magnetic junctions which are based on the tunnel magnetoresistance effect. As the name suggests a tunnel magnetic junction is based on the quantum mechanical tunnelling of electrons across a narrow insulating layer separating two ferromagnetic layers. Again, there is a higher resistance when the magnetisations in the ferromagnetic layers are antiparallel.

7.6 Closing remarks

In this chapter we have looked at a small selection of properties of materials at the nanoscale that arise from the quantum behaviour of electrons. These are well established areas of nanoscience, which have led to major advances in technology. There are plenty of others. They have all been made possible through ingenious ways of synthesising materials at these minute length scales in remarkably controlled and reproducible ways. There have been spectacular advances in the observation and characterisation of materials down to the atomic scale. Materials science at the nanoscale has also benefited since the late 1980s from the availability of software to model materials at the atomic scale based on quantum mechanics, principally density functional theory (DFT).

Further reading

Wolf, E L, *Nanophysics and Nanotechnology*, 2nd edition, Wiley-VCH (2009).

Benelmekki, M, *Nanomaterials: The Original Product of Nanotechnology (IOP Concise Physics)*, Morgan Claypool (2019).

Stevens, S Y, Sutherland, L M and Krajcik, J S, *The Big Ideas of Nanoscale Science and Engineering – A guidebook for secondary teachers*, National Science Teachers Association (2009).

8
Collective behaviour

The ability to reduce everything to simple fundamental laws does not imply the ability to start from those laws and reconstruct the Universe... The constructionist hypothesis breaks down when confronted with the twin difficulties of scale and complexity. The behaviour of large and complex aggregates of elementary particles, it turns out, is not to be understood in terms of a simple extrapolation of the properties of a few particles. Instead at each level of complexity entirely new properties appear, and the understanding of the new behaviours requires research which I think is as fundamental in its nature as any other... Psychology is not applied biology, nor is biology applied chemistry... The whole becomes not merely more, but very different from the sum of its parts.

From Anderson, P W, *More is different*, Science **177**, 393 (1972). Reprinted with permission from AAAS.

8.1 Concept

The science of materials spans length and time scales. At each larger scale new, fundamental science emerges from the collective behaviour of particles or other entities at smaller scales.

8.2 More is different

One of the dominant trends in materials science over the past fifty years has been the study of ever smaller samples. This has led to the spectacular rise of nanoscience – the study of materials at the sub-micron scale. A host of experimental techniques has been developed to examine the structures and chemical bonding of materials at the atomic scale. At the same time advances in theory, software and computers have enabled these features to be modelled atomistically using only the fundamental laws of quantum mechanics and electrodynamics. The ability of these experimental and computational methods to study materials at the same length scales has led to remarkable synergy and advances in the nanoscience of materials, as discussed in Chapter 7.

This reductionist approach to materials science has been frequently applied to understanding properties and processes in *macroscopic* materials. It appears to stem from a belief that it is only by studying materials at the atomic scale that a complete understanding of processes and properties can be achieved[1].

Many properties and processes in materials involve physics across a range of length scales. At each length scale new physics arises. A good materials scientist takes a holistic view, looking across all the relevant length scales. A principal goal is to understand how the physics at one scale influences the physics at other scales.

One of the most significant features of a large group of atoms, say a quintillion (10^{18}) atoms, is the emergence of new properties and processes that are not displayed by a small group. A property of a system is emergent if it is not displayed by the parts the system comprises. The spontaneous electric polarisation of a ferroelectric crystal is an emergent property of the crystal because the constituent atoms are not polarised when they are isolated. At a larger scale, the electrostatic and elastic energies of a ferroelectric crystal become important. Their minimisation leads to the emergence of domains separated by domain boundaries where the polarisation changes direction. The mobility of the domain boundaries controls the ease with which the overall polarisation of the crystal can change in response to an applied electric field. The emergent properties and processes at each scale introduce new physics.

Consider a crystalline material in an open thermodynamic system where the intensive variables of temperature, pressure and chemical potentials are defined. Some properties of the crystal depend only on the positions and atomic numbers of the relatively small number of atoms in a repeat cell of the crystal lattice, together with the structure of the lattice. Such properties include chemical bonding and cohesive energy, elastic constants, atomic vibration spectra, spontaneous electric and magnetic polarisations, electric and magnetic susceptibilities, magnetostriction and electrostriction constants, optical absorption and emission and so on. They are all emergent properties because they are not properties of the constituent atoms by themselves. They are determined by the fundamental laws of quantum mechanics and electrodynamics applied to the relatively small group of atoms in a cell that repeats throughout space.

There are also processes in crystals in which a range of length scales is involved, with new physics arising at each length scale. Examples include permanent (plastic) deformation, creep, irradiation damage, phase changes, sintering, solidification and casting, recrystallisation and recovery processes, corrosion, tribology, fracture and fatigue. These processes affect properties such as hardness and strength, brittleness and ductility, formability and durability, electrical and thermal conductivities and thermodynamic stability. They all involve crystal defects. Atomic interactions are still relevant because, for example, they govern the atomic structures and intrinsic mobilities of defects. The new physics includes elasticity, diffusion and long range electrostatic fields which describe the interactions and migration of defects over distances much greater than the atomic scale.

Long-range interactions between defects and with external loads are mediated by elastic fields. In insulating materials charged defects also interact electrostatically. The theory of elastic interactions between defects is an example of the emergence of new physics from the collective behaviour of many atoms. Although in principle

---

[1]One area where this belief is justified, at least to some extent, is the kinetic theory of gases, where statistical physics is applied to the atomic or molecular constituents of gases to explain their macroscopic properties. But even in gases new physics arises at larger length scales, such as turbulence.

elastic interactions between defects are governed by the same fundamental laws of quantum mechanics and electrodynamics, it would be a senseless task to model these interactions in a quintillion atoms using these fundamental laws alone, even if it could be done. It would also be extremely difficult to extract the new physics from the results. Thinking in terms of only the fundamental laws buries the new physics under a mass of irrelevant detail. Instead the theory of elasticity provides lucid and accurate descriptions of long-range interactions between defects, often without the need for a computer.

In the following section three examples of familiar processes in materials are discussed in which physics at the atomic scale is always important, but a complete understanding involves physics at larger length scales.

8.3   Three examples of processes involving multiple length scales

8.3.1   Electronic conduction

Although aluminium is a very good electrical conductor at room temperature its resistivity is not zero. As discussed in Section 3.5 conduction electrons are scattered by atomic vibrations and structural defects. Between these scattering events conduction electrons travel at a speed of around 1% of the speed of light. The average distance travelled between successive scattering events is called the mean free path, and it decreases as the temperature increases owing to the increased probability of being scattered by atomic vibrations. Conduction electrons are also scattered by defects such as vacancies, impurities, dislocations and grain boundaries. As a result of this scattering conduction electrons travel along the wire in a *diffusive* manner, with a drift speed that is many orders of magnitude smaller than the speed between scattering events. This is the physics underlying Ohm's law, and we see it emerges only in sufficiently large samples where all these scattering mechanisms are represented.

It is possible to make a metallic wire much shorter and thinner and to connect it to electrodes using the techniques of nanotechnology. At the same time the temperature can be reduced to cryogenic levels. The length of the wire can be made significantly less than the mean free path for scattering by atomic vibrations. Now the conduction electrons are able to travel through the wire at about 1% of the speed of light, which has been described as *ballistic* transport. The mechanism of their transport changes from the classical physics of Ohm's law to the quantum physics of treating the conduction electron as a wave. When the wire is atomically smooth and there are no internal defects such as vacancies it behaves like a wave-guide. Its conductance is determined by the number of transverse modes that can fit into the narrowest part of the wire, and that the current-carrying electrons can access at the Fermi energy. Each open channel contributes $2e^2/h = 7.75 \times 10^{-5}\Omega^{-1}$ to the conductance, where e is the charge on the electron and h is the Planck constant. The conductance has become quantised and independent of the metal. When electrons are transported ballistically the concepts of resistivity and Ohm's law no longer apply.

Electronic conduction is a defining property of a metal. But it arises only when the metal is in a condensed state, i.e. a solid or liquid. It seems obvious that when aluminium atoms are separated by large distances, as in a dilute vapour phase, electrons cannot be transferred from one atom to another as they would in a condensed state. The reason is simple energetics. An outer electron of each aluminium atom is bound by the first ionisation energy, which is $578 \, \text{kJ mol}^{-1}$, or $5.99 \, \text{eV}$ per atom. If an electron is added to a neutral aluminium atom $43 \, \text{kJ mol}^{-1}$, or $0.45 \, \text{eV}$ per atom, is released. Therefore to transfer an electron between two well separated neutral aluminium atoms requires $535 \, \text{kJ mol}^{-1}$, or $5.54 \, \text{eV}$ per electron. That is far more energy than is available thermally under normal laboratory conditions, so it does not happen.

When aluminium crystallises it is a good conductor. At room temperature and pressure the atoms in a crystal of aluminium are approximately $0.286 \, \text{nm}$ apart. Consider a thought experiment in which we expand the crystal lattice of aluminium at a steady rate so that the separation between nearest neighbours increases steadily from $0.286 \, \text{nm}$ to one metre. There can be no doubt that when we reach the separation of one metre the aluminium crystal is no longer able to conduct electricity. At some point in this process the crystal undergoes a transition from a metal to an insulator. The transition occurs remarkably abruptly at a particular separation of the atoms. As the atomic separation increases the electrons that were originally able to move freely from atom to atom become increasingly localised at ion cores. The barrier to transferring an electron from one atom to another increases in width and height. As long as the barrier is not too wide the transfer can occur by quantum tunnelling[2] through the barrier. This is a cooperative process: when electrons can tunnel they help to screen the interaction between other electrons and their ion cores. The amount of screening depends on the ease of electron tunnelling between atoms, and the ease of tunnelling depends on the amount of screening. This cooperative behaviour results in an abrupt transition from metal to insulator as the crystal is dilated.

Fig. 8.1 shows an abrupt metal-insulator transition in silicon as a function of doping with phosphorus at a temperature of $1 \, \text{mK}$. Phosphorus has five valence electrons. When it occupies a substitutional site in silicon only four of them are taken up in bonds to the four neighbouring silicon atoms. At low temperatures the fifth electron enters a hydrogenic-like orbit with a radius of about $2.5 \, \text{nm}$ and its binding energy to the P^+ ion core is about $0.045 \, \text{eV}$. At a temperature of $1 \, \text{mK}$ the fifth electron is bound to the ion core. At a concentration of $3.8 \times 10^{18} \, \text{cm}^{-3}$ there is an abrupt transition from insulating to metallic behaviour. At this concentration the hydrogenic orbits of neighbouring P atoms overlap. An electron can then tunnel from one ion to the next and screen the interaction between the electron and the P^+ ion core, about which it orbits, liberating the orbiting electron. The more that are liberated the greater the screening and so on.

The thought experiment explains why some materials, notably transition metal oxides, that are predicted to be metallic are in fact insulating. Standard 'band theory'[3] predicts they are metallic because it does not take into account the energy required to transfer an electron from one atom to another. In some cases this energy is negligible because the screening is effective and the theory correctly predicts the material is a metal. But in other cases, where the atomic orbitals do not overlap sufficiently to enable effective screening, the theory incorrectly predicts the material is a metal when it is in fact a good insulator.

---

[2]Quantum tunnelling is described in sections 6.6 and 6.8.
[3]Band theory is discussed in section 6.5.

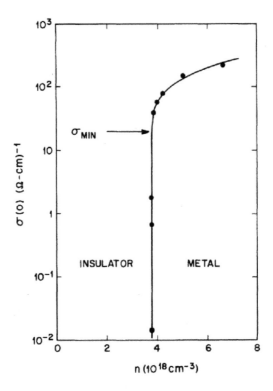

Fig. 8.1 Abrupt metal-insulator transition in silicon as a function of doping with phosphorus at a temperature of 1 mK. The graph shows the electrical conductivity as a function of the phosphorus concentration. Reprinted figure with permission from Rosenbaum, T F, Andres, K, Thomas, G A and Bhatt R N, Phys. Rev. Lett. **45**, 1723 (1980). Copyright (1980) by the American Physical Society.

This example illustrates the influence of length scale on the mechanisms of electronic conduction in a metal, and even on whether the material is a metal or insulator.

8.3.2 Plasticity

In Section 4.4 dislocations were identified as the agents of plastic or irreversible deformation. The elastic fields of dislocations enable them to interact with each other over distances that extend throughout a macroscopic sample. In a cubic centimetre of heavily deformed metal there can be 10^{12} cm of dislocation lines, which is equivalent to 10^{12} dislocations threading through a square centimetre. That amounts to ten million kilometres of dislocation line in one cubic centimetre of heavily deformed metal. If they were all laid out end to end they would reach to the moon and back thirteen times. During plastic deformation they organise themselves into cellular structures, with cell walls made of dislocation tangles and cell interiors relatively free of dislocations. Dislocations also interact with each other at short range forming obstacles to further slip, which gives rise to work hardening. In work hardening the stress required to sustain plastic deformation increases with the amount of plastic deformation. Particles of other phases present in the material act as obstacles to the motion of dislocations because the dislocations either have to loop around them or cut through them.

A particularly interesting idea has been advanced by L M Brown[4] to understand the work-hardened state of a metal. He argues it reaches a self-organised critical state, where avalanches of local slip can take place anywhere in the sample, punctuated by periods where the strain increases without any dislocation motion until the system again becomes critical. This state leads him to describe plastic deformation as 'constantly intermittent'. Similar ideas have been put forward to describe earthquakes, and there are some obvious resemblances.

The story of the generation and interaction of dislocations in plastic deformation is written in the language of elasticity as well as atomic interactions. The theory of elastic interactions between defects is a paragon of the emergence of new physics from the collective behaviour of large numbers of atoms. The intermittent nature of plastic deformation resulting from the self-organisation of dislocations into a critical state is also new physics that emerges only at larger length scales. The atomic scale is relevant because it determines the intrinsic mobilities of dislocations, their interactions at short range, and other important factors we have not mentioned such as the possible existence of metastable faults in the slip plane. The multi-scale nature of plasticity involving the collective behaviour of huge numbers of flexible, mobile, interacting line defects has made it one of the most difficult and least understood phenomena in the whole of classical physics.

8.3.3 Fracture

Fracture is the breaking of a sample or component into two or more parts. There are diverse mechanisms but they generally involve the nucleation and growth of cracks. The physics of fracture spans multiple length scales.

Purely brittle fracture is an idealisation because it is very slow, reversible and there is no energy dissipation into heat. Consider a body with a constant load applied to its exterior. In purely brittle fracture the energy required to make the crack grow has two components. The first is the energy of the two new surfaces that are formed. The second is the increase in the elastic energy of the body due to growth of the crack. At a constant applied load the elastic energy of the body increases as the crack grows because the body becomes more compliant. The sum of these two energies has to be supplied by the work done by the external loading mechanism, otherwise the crack shrinks by closing. This is Griffith's criterion for growth of a brittle crack. We see that it involves the elastic energy of the entire body, the potential energy of the external loading mechanism and the creation of crack surfaces by breaking atomic bonds at the crack tip. It is evidently multiscale.

In reality fracture is almost always accompanied by some degree of plasticity and energy dissipation, and so crack growth is almost always irreversible and it consumes more energy than a purely brittle fracture. The stresses near a loaded crack tip are much greater than those far from the crack. In crystalline materials dislocation sources

[4]Brown, L M, Philosophical Magazine **96**, 2696 (2016).

near the crack may be activated, and dislocations may be emitted from the crack tip. Some dislocations, called 'shielding dislocations', move away from the crack tip and screen the applied load so that bonds at the crack tip are less stressed. The work required to fracture the material increases as a result of this dislocation activity. The local forces acting on interatomic bonds at the crack tip are dependent on dislocation activity ahead of the crack as well as the external loading. Now the physics of crack growth involves the length scale associated with dislocation activity ahead of the crack.

In ductile metals fracture occurs by the nucleation and growth of voids ahead of a crack. The crack grows by linking up with the voids. The voids can nucleate on particles of other phases by decohesion of the interface surrounding the particles. In ductile pure metals voids nucleate at other stress concentrations such as regions of high dislocation density.

Some crystalline materials undergo a transition with increasing temperature from brittle behaviour at low temperatures to ductile behaviour at higher temperatures. This brittle to ductile transition occurs in crystals where dislocation motion is thermally activated, and the transition is associated with the activation of dislocation sources near the crack tip and at the crack tip at the transition temperature. In some materials, such as silicon, the transition is very abrupt because there are relatively few dislocation sources in the material. The sinking of *The Titanic* was a consequence of a ductile to brittle transition of its steel hull[5]. In the icy waters of the North Atlantic the steel became brittle. When the ship struck an iceberg the hull fractured rather than absorbing the energy of the impact through plastic deformation.

In polycrystalline materials cracks may propagate along grain boundaries rather than through grain interiors. Atoms of some elements may promote grain boundary fracture by segregating to the boundaries and weakening the bonding across the boundary. In steels, segregation of antimony, tin, sulphur and phosphorus may lead to such intergranular embrittlement. In contrast, carbon and boron in steels enhance cohesion at grain boundaries. An understanding of why some elements reduce cohesion at grain boundaries while others enhance it involves the changes in the atomic and electronic structures of grain boundaries these elements bring about. The ease with which bonds are broken and reformed with neighbouring atoms may be as important as the strength of bonds because the former are involved in dislocation emission.

We have already mentioned that a void may be nucleated at a particle as a result of decohesion taking place at the interface of the particle. More generally, cracks are nucleated at stress concentrations, such as dislocation pile-ups at grain boundaries.

The physics of fracture is multi-scale and what happens at one scale directly affects what happens at other length scales. Dislocation activity ahead of cracks affects the local loading of atomic bonds at the crack tip. Segregation of impurities at the crack tip affects the extent of dislocation activity ahead of the crack because it affects the stresses that can be sustained at the crack tip and hence the stresses ahead of the crack. Almost all the prevailing models of fracture consider only one length scale, and cannot be considered complete.

---

[5] Felkins, K, Leighly, H P. Jankovic, A, JOM **50**, 12 (1998).

Further reading

Cottrell, A H, *An Introduction to metallurgy*, 2nd edition, The Institute of Materials (1995).

Ghoniem, N and Walgraef, D, *Instabilities and self-organization in materials*, Volumes 1 and 2, Oxford University Press (2008).

Peierls, R E, *The laws of nature*, George Allen & Unwin Ltd. (1955).

Sutton, A P, *Physics of elasticity and crystal defects*, Oxford University Press (2020).

9
Materials by design

There is often frustration from industry when a new material is reported in the literature to have attractive properties for aerospace based on limited testing and, in particular, limited knowledge of the end application. In reality there is a huge gap between such early-stage research and the formidable requirements to certify a new material for use. Likewise, funding agencies will often show interest in exciting new materials. For some material applications, such as electronic devices, time from invention to application is relatively short compared with materials for jet engines, which can take many years.

David Rugg, Senior Engineering Fellow in Materials at Rolls-Royce plc, interviewed by John Plummer in *Understanding a way to fly high*, Nature Materials **15**, 820 (2016). Reprinted by permission from Springer Nature, Copyright © 2016.

9.1 Concept

To design a material is to determine the optimum chemical composition, internal structure and method of fabrication to meet the requirements for a particular application.

9.2 Introduction

The process of designing a material starts with a good understanding of the requirements of the intended application. The designer recognises how the properties of the material are controlled by its structure over a range of *length scales*, and how the structure is determined by the method of fabrication. The designer also considers the *time scales* associated with possible changes to the material during its life in service. Many aspects of materials science are involved in the design and fabrication process.

Materials design does not necessarily involve the creation of a new material. It may be that an existing material may be improved by a relatively small adjustment to its composition, surface treatment or method of fabrication. Sometimes this is all that is needed to give the manufacturer a competitive edge. It is often a more attractive outcome to a manufacturer than the replacement with a completely different material because it is usually cheaper and it usually requires less testing and development. However, to improve an existing material requires an *insight* into why it is not optimal and how it can be improved.

Materials design should not be confused with materials selection and materials discovery. Materials selection involves making a choice from a range of existing materials for an intended application. It may be that none of the available materials suit the application.

Materials discovery is a relatively recent term for the application of techniques of data science and artificial intelligence to create, curate and search data on materials for candidates with potentially useful properties. It also uses 'high throughput' techniques of combinatorial chemistry and computational materials science to search quickly a range of materials for desirable properties. Materials discovery is the reverse of materials design. Instead of a seeking to create a suitable material for a particular application, the *modus operandi* of materials discovery is to seek applications for a material that has promising properties. The quotation above from Rugg's interview expresses a frustration about materials discovery often heard from manufacturing industries, not only aerospace.

There are Nobel prize winning discoveries of remarkable new materials that have had limited applications in technology decades after their discovery. Evidently, the discovery of a new material with exceptional properties does not necessarily guarantee it will create a disruptive technology[1]. Some other Nobel prize winning discoveries of new materials, such as the ceramic superconductors, have led to disruptive technologies, but only after decades of research to develop the materials for applications. One reason for the delay in the case of the ceramic superconductors is that they are brittle and it was not trivial to make electromagnets from them. They show there may be unavoidable reasons for the delay between discovery and deployment of a new material in technology. It is often true that the financial cost of the discovery of a material is a small fraction of the development cost. In contrast, a good design reduces the time and cost of development. Since materials design starts from a full understanding of the requirements of a particular application, it is targeted and the material is more likely to be used in the application for which it was designed.

It may seem obvious that the development of a new material to replace an existing material requires a complete appreciation of the requirements of the intended application if it is to be adopted. The frustration expressed by Rugg is a reflection of how rarely this appreciation exists when a new material is announced. This may explain the apparent lack of interest shown by manufacturing industries, at least in Europe, in materials discovery and data-driven materials science more generally, that has been noted in a recent review article[2].

9.3 Microstructure

A central aspect of materials design is the concept of microstructure, aspects of which we have already encountered in Chapters 4 and 8. It includes structural defects such as grain boundaries, networks of dislocations and clusters of point defects, as well as particles of other phases, microcracks and variations of composition in solid solution.

---

[1] A disruptive technology brings about a step-change to the way that consumers, industries or businesses operate. If there were earlier related technologies they are made redundant by the superior qualities of the disruptive technology.

[2] Himanen, L, Geurts, A, Stuart Foster, A and Rinke, P, *Advanced Science* **6**, 1900808 (2019).

These features appear at length scales much larger than the atomic scale but much smaller than the macroscopic scale of an engineering component. They may have a controlling influence over the mechanical, electrical, magnetic and optical properties of a material. Their presence illustrates the emergence of new physics resulting from collective interactions between atoms. Microstructure is normally an indication that the material is not in thermodynamic equilibrium, but in some metastable or unstable state. There is a range of time scales associated with the relaxation of different aspects of the microstructure towards equilibrium. Many of these relaxation processes involve thermal activation, and the rates at which they occur depend sensitively on the temperatures to which the material is exposed during manufacture and in service. This is one reason why the method of fabrication and service conditions must be taken into account in the design of the material. Another reason is that the mechanical loading to which the material is subjected during fabrication and in service also affects the populations of microstructural features.

Some materials are driven away from thermodynamic equilibrium by the conditions of their service. Examples include materials constantly subjected to irradiation and varying stress, electric or magnetic fields. In these cases the microstructure of the material also evolves, leading to significant changes in properties and possibly even failure. This is yet another reason why it is important in materials design to take into account the service conditions.

Microstructural features may interact with each other creating further collective behaviours, such as the reduction of dislocation densities by the migration of grain boundaries during recrystallisation, or the nucleation of cracks at stress concentrations caused by dislocation pileups. As discussed in Chapter 1 the state of equilibrium in a material is affected by thermal, mechanical and chemical interactions with its environment. Physicists use the term 'complex system' to describe the emergence of collective behaviours arising from interactions between the various parts of a system, and how the system interacts with its environment. Materials display all the aspects of a complex system.

The design of a material is most effective if it is treated as a complex system using what engineers call a 'systems approach'. As before, the process begins by enumerating the properties the material must display to meet the requirements of its intended application. The required properties determine the class of material (metal, ceramic, polymer, composite etc.), its composition, microstructure and atomic structure. This determination involves modelling the material at different length scales, from the atomic scale to the microstructural and macroscopic length scales with information flowing up and down the length-scale hierarchy to integrate the models. The required structure, at all length scales, determines how the material is to be fabricated. This is guided by modelling the fabrication process. After the material is fabricated it is tested to see whether it has the required properties. If necessary, the properties of the material can be adjusted by altering the composition of the material and its fabrication process, again guided by modelling. The evolution of the microstructure of the material in service can also be modelled. If necessary the whole process can be further adjusted to ensure the required properties are maintained throughout the intended service life. This approach to the design and fabrication of materials is known as 'Integrated Computational Materials Engineering' (ICME), and it was developed by Gregory Olson at Northwestern University[3].

9.4   An example: replacing the 'nickel'

A recent example of materials design is work[4] at the National Institute of Standards and Technology (NIST) in the US by Eric Lass and colleagues. They were asked by the US Mint to find a new alloy to replace the copper-nickel alloy used to make five cent coins, known as 'nickels'.

Since 2006 the cost of making five cent coins had been more than five cents because the cost of nickel had risen so much. The goal was to reduce the cost by 40%. In a materials discovery approach one would create a promising material and design a product based on the strengths and limitations of the material. The success of the work by Lass *et al.* stems from the reversal of this process: 'the researchers began with a list of needs from the Mint and designed a material to meet the product need'[5]. This was a strategy of materials design rather than materials discovery. The researchers began by understanding the requirements of the replacement alloy as stipulated by the US Mint. The replacement alloy had to keep the following properties the same as in the previous five cent coin: electrical conductivity, density, colour, yield strength, work hardening coefficient, corrosion resistance and toxicity. These requirements enabled the same equipment to be used to manufacture the coins in the mint, and to enable existing vending machines to recognise them as five cent coins. The cost of retooling a manufacturing process is often prohibitive and the requirement that the same manufacturing equipment can be used with a new material is not uncommon. The coins must also be able to withstand the wear they normally suffer during their 30 years in circulation. They must be non-toxic, antifungal, and the constituent materials must not leach out of them. They must also resist corrosion and tarnishing as coins are exposed to a variety of environments including human sweat. The new coins had to be recyclable by melting them down and reusing them. An alloy that does not meet all these requirements is unacceptable to the US Mint.

Lass *et al.* used the systems approach of ICME to design and fabricate three prototype alloys in the Cu-Ni-Zn-Mn system that met the design objectives. The design approach recognised that some of the design objectives may be conflicting and it was able to optimise the composition of each alloy and its fabrication process to satisfy them all. The final solutions were also robust to small variations in their compositions and their production process.

9.5   Self-assembly

Self-assembly and self-organisation describe the autonomous formation of patterns or structures resulting from short-range interactions between separate components that

---

[3]Olson, G B, Science **277**, 1237 (1997).

[4]Lass, E A, Stoudt, M R and Campbell, C E, Integrating Materials and Manufacturing Innovation **7**, 52 (2018).

[5]Gillespie, A, *Materials by design cooking up innovations with the Materials Genome Initiative.* https://tinyurl.com/y7bj7255

Fig. 9.1 Image of a bubble raft formed by the self-assembly of bubbles. There is a small angle grain boundary from left to right made up of edge dislocations.

start from a disordered configuration. Although the expressions are frequently used interchangeably it is useful to distinguish between them. Self-assembly is a process driven by a reduction of free energy – it is a process of equilibration. Thus, crystallisation is a process of self-assembly. Self-organisation is an energy consuming process. Examples are living cells, dislocation cell structures in plastically deformed metals, and murmurations of starlings. Self-organisation in 'active materials' as models of living matter is discussed in Chapter 11. In this section I will focus on self-assembly. In order for components to self-assemble into an ordered structure they have to be mobile. This is why self-assembly usually takes place in fluids or on surfaces, where the components move by Brownian motion if they are sufficiently small.

9.5.1   The bubble raft

A beautiful illustration of self-assembly in two dimensions is the bubble raft experiment of Bragg and Nye[6]. In this experiment bubbles about 1 mm in diameter are introduced into a tray containing a dilute soap solution. The bubbles are generated by passing air at a steady rate through a fine nozzle placed beneath the surface of the solution. As soon as the bubbles reach the surface of the solution they attract each other and self-assemble into rafts. Within a raft the bubbles are closely packed forming a hexagonal arrangement.

Figure 9.1 shows a bubble raft in which there is a small angle grain boundary from left to right across the middle of the photograph. The grain boundary consists of an array of edge dislocations. The bubble raft is an excellent model of a crystal in two dimensions. Apart from edge dislocations it can display point defects such as vacancies (missing bubbles), interstitial defects (undersize bubbles), substitutional defects (oversize bubbles) and grain boundaries. By agitating the bath the model can be made dynamic, with mobile dislocations and grain boundaries.

The self-assembly of the bubbles is a consequence of forces acting between them. The force of interaction between two bubbles is similar to that shown in Fig. 3.3b.

[6]Bragg, W L and Nye, J F, Proc. R. Soc. Lond. A **190**, 474 (1947).

The force is one of attraction when the bubbles are separated[7]. When the bubbles are in contact the force of attraction is balanced by a short-range force of repulsion[8].

9.5.2   Photonic crystals

A technologically useful example of self-assembly is the crystallisation of colloidal structures to make photonic crystals[9]. Photonic crystals are periodic structures like ordinary crystals except that the 'atoms' are particles with sizes comparable to the wavelengths of light. The particles may contain billions of atoms.

In Section 6.5 we outlined band theory, where electronic states in crystals are in bands of energy. The bands of electronic states are often separated by gaps where there are no electronic states. An electron with an energy in one of the gaps cannot travel through the crystal. Similar behaviour is observed in a photonic crystal. There are bands of states where photons can travel through the crystal separated by gaps where photons cannot propagate. One dimensional photonic crystals, which comprise stacks of thin layers of insulating materials with alternating refractive indices, are used for low and high reflective coatings on lenses and mirrors. There are also naturally occurring photonic crystals. The iridescence of opals and the colours of some butterfly wings are due to photonic crystals.

Particles with narrow size distributions are described as 'monodispersed'. To create a photonic crystal the size distribution has to be as narrow as possible. Otherwise there are particles that behave as point defects destroying the periodicity of the crystal, and possibly introducing photon states inside the band gap.

The most commonly used particles are spherical, and they generally create close-packed crystal structures. Monodisperse silicon dioxide spheres can be fabricated with sizes ranging from a few nanometres to a few micrometres. Other materials that have been fabricated as monodisperse spheres include titanium dioxide, zinc oxide, zinc sulfide, cadmium sulfide, zinc selenide, lanthanide compounds, gold, silver, lead, bismuth, antimony and tellurium[9]. In addition it is possible to synthesise core-shell particles in which the core of a particle is coated by another material forming a shell. These particles have found applications in photonics, catalysis, sensing, and imaging and targeted destruction of cancer cells[10]. A particular case of a core-shell particle is a hollow sphere where the core is removed chemically after the shell has been deposited. In this way hollow metallic, semiconducting and insulating spheres have been produced.

Three dimensional crystals of spherical particles of silica may be formed by suspending them in a solvent such as water or ethanol[11]. A vertical substrate is withdrawn from the solution at a constant rate. Crystals of between two and several hundred close-packed layers of silica spheres self-assemble in the meniscus formed on the substrate. As the solution evaporates the crystal consists of a stable close-packed configuration of silica spheres in air.

[7]Nicolson, M M, Mathematical Proceedings of the Cambridge Philosophical Society **45**, 288 (1949).

[8]Lomer, W M, Mathematical Proceedings of the Cambridge Philosophical Society **45**, 660 (1949).

[9]Galisteo-López, J F, Ibisate, M, Sapienza, R, Froufe-Pérez, L S, Blanco, Á and López, C, Advanced Materials **23**, 30 (2011).

[10]Loo, C, Lowery, A, Halas, N, West, J and Drezek, R, Nano Lett. **5**, 709 (2005).

[11]Jiang, P, Bertone, J F, Hwang, K S and Colvin, V L, Chemistry of Materials **11**, 2132 (1999).

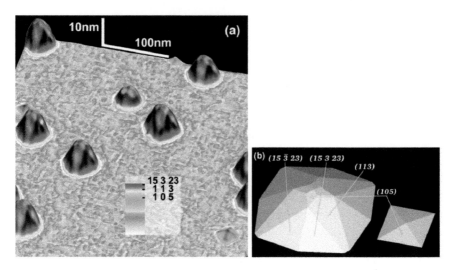

Fig. 9.2 (a) Scanning tunnelling micrograph of germanium/silicon pyramids (green) and domes (multi-coloured) coexisting on a silicon (001) surface. The colouring indicates the orientations of the different faces of the islands, differentiating the {105} facets of the pyramids from the higher index facets of the domes. The inset shows the facet colour coding. The islands self-assembled through diffusion of germanium atoms deposited by molecular beam epitaxy on the silicon surface at 3 monolayers per minute at 600°C to give an equivalent coverage of 8 monolayers. (b) An idealised representation of the dome and pyramid at their true aspect ratio with representative facets labeled. From Rudd *et al.*[12]. Copyright © American Scientific Publishers 2007.

9.5.3   Quantum dots

Quantum dots were introduced in Section 7.3. In this section I illustrate the self-assembly of quantum dots with the example of germanium quantum dots on silicon surfaces[12]. These are grown by depositing germanium atoms onto a silicon (001) surface[13] under ultra high vacuum conditions using molecular beam epitaxy. Silicon and germanium share the same diamond cubic crystal structure, but the lattice constant of germanium is about 4% larger than that of silicon. The first germanium atoms to arrive on the surface continue the crystal structure of the silicon by forming a thin layer, called a wetting layer. The wetting layer consists of about three atomic layers of germanium. But this puts the germanium under a large compressive strain. When more germanium atoms arrive on the surface islands start to form in which the strain is reduced, but not eliminated completely. The reduction in the elastic strain energy more than compensates for the increase in total surface energy when islands are formed. When the islands are small they have the shape of pyramids while larger islands are more rounded structures called 'domes', as illustrated in Fig. 9.2.

---

[12]Rudd, R E, Briggs, G A D, Sutton, A P, Medeiros-Ribeiro, G and Williams, R S, Journal of Computational and Theoretical Nanoscience **4**, 335 (2007).

[13]The (001) surface is parallel to a face of the cubic unit cell.

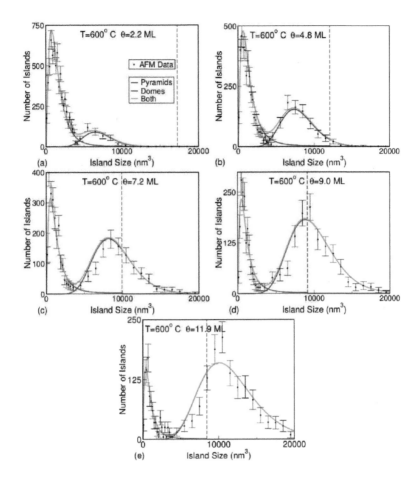

Fig. 9.3 Island size distributions of germanium quantum dots on silicon (001). The solid curves are the distributions calculated with the model based on the assumption of equilibrium. The dots with error bars show the results of experimental measurements using atomic force microscopy (AFM). The curves give the numbers of islands of each type and the total number. θ is the amount of germanium deposited in equivalent monolayers (ML). Note the vertical scales are different. The bimodal size distribution is characteristic of two island types. The broken vertical line indicates where the model distribution for pyramids was truncated. From Rudd *et al.*[12]. Copyright © American Scientific Publishers 2007.

A bimodal size distribution is observed of smaller islands (pyramids) and larger islands (domes), as illustrated in Fig. 9.3. Bimodal distributions of quantum dots are found in other systems too, e.g. gallium nitride islands grown on aluminium nitride[14].

The amount of germanium deposited on the silicon surface is called the 'coverage'. The distributions of the island sizes depend on the coverage and the temperature of the

[14] Adelmann, C, Daudin, B, Oliver, R A, Briggs, G A D and Rudd R E, Phys. Rev. B **70**, 125427 (2004).

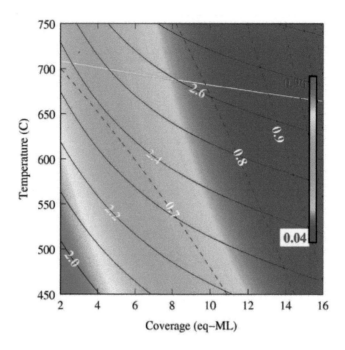

Fig. 9.4 The nanostructure diagram for germanium islands on the silicon (001) surface. It first appeared in Rudd, R E, Briggs, G A D, Sutton, A P, Medeiros-Ribeiro, G and Williams, R S, Phys. Rev. Lett. **90**, 146101 (2003), from where it is reprinted with permission. Copyright (2003) by the American Physical Society. The colour scale indicates the fraction of islands that are pyramids (from 4% to 96%). The black contours indicate distribution widths relative to the mean island size, with the solid black curves indicating the distribution widths for pyramids and the dashed curves those for domes.

deposition process. There are two possible explanations for this behaviour. The first is that the populations of the pyramids and domes are controlled by a kinetic process called 'Ostwald ripening'. This is a process in which larger islands grow and smaller islands shrink and eventually disappear by surface diffusion. The second is that there is an equilibrium population of both pyramids and domes for each temperature and coverage. There is experimental evidence that supports the idea of an equilibrium state, or at least a metastable state. After long anneals at 550°C the island distributions reached a stationary state. Islands of significantly disparate sizes were observed next to each other. Both observations suggest that Ostwald ripening was not taking place.

Rudd *et al.*[12] presented a model of the distributions of the sizes of the two islands as a function of temperature and coverage of germanium on the silicon surface. The model assumed the distributions are governed by equilibrium. Equilibrium requires that the chemical potential of germanium is the same in all islands regardless of their type and size. Since the islands are small equilibrium fluctuations in their sizes are significant. This is the fundamental reason why there are *distributions* of island sizes at a given temperature and coverage. To treat distributions at equilibrium the model used statistical mechanics. The model calculates the equilibrium populations of domes and pyramids, their sizes and the widths of their distributions. The results are conveniently summarised in a map called a 'nanostructure diagram' with axes temperature and coverage. It is shown in Fig. 9.4.

The nanostructure diagram does for nanostructures what a temperature-composition phase diagram does for macroscopic alloys. We see in Fig. 9.4 that at relatively low temperatures and coverages pyramids dominate and at high temperatures and coverages domes prevail. If we select a temperature of 600°C the islands are primarily pyramids at 2.2 equivalent-ML coverage, as we see also in Fig. 9.3a. With increasing coverage at 600°C we move along a line parallel to the coverage axis in the nanostructure diagram. When the coverage reaches 4.8 equivalent-ML the distributions are as shown in Fig. 9.3b. On the nanostructure diagram we are in the middle of the green region indicating that approximately half the islands are pyramids. The widths of the distributions of sizes of the pyramids and domes, relative to their mean values, can also be estimated from the solid and broken black contours. On increasing the coverage further more pyramids are converted into domes. Narrow widths are preferable for applications, and the dome distributions are narrower than the pyramid distributions throughout the diagram.

9.6   Smart materials

Smart materials respond to changes in their environment by altering one or more properties in a reversible manner. A familiar example is the switch on a domestic kettle which turns off the power automatically when the water boils. The switch contains a metallic alloy which has one shape below the boiling point of water and another shape above. It displays the 'shape-memory' effect, the origin of which is a particular kind of phase change in the material called a martensitic transformation. Similar reversible changes in the shape of an object resulting from martensitic phase changes can be induced by stresses and magnetic fields. Another familiar example is a solar cell, where a voltage is created by shining light on the device. The origin of the effect, which is known as the photovoltaic effect, is the generation of excitons by the incident photons, and the separation of electrons and holes by internal electric fields in the device. There is a large range of effects in materials that are exploited in the design of smart materials. They include:

- The thermochromic effect where the material changes colour in response to changes of temperature.
- The halochromic effect where the material changes colour in response to changes in the acidity of the environment.
- The photochromic effect where the material changes colour in response to light.
- The photomechanical effect where the material changes shape in response to light.
- The piezoelectric effect where a voltage is created when a stress is applied to the material, and conversely a strain is generated when a voltage is applied.

- The magnetostrictive effect where a material changes shape when a magnetic field is applied, and conversely the magnetisation in the material changes when a stress is applied.
- The electrostrictive effect where a material changes shape when an electric field is applied and conversely the electric polarisation in the material changes when a stress is applied.
- The thermoelectric effect where a voltage is created when the material is placed in a temperature gradient, and conversely the temperature of the material changes when an electric field is applied.
- The magnetocaloric effect where the temperature of the material changes when it is exposed to a changing magnetic field.

9.6.1   Self-healing materials

These are smart materials that can automatically repair damage to themselves without human detection of the damage or intervention to repair it.

Concrete is the most widely used man-made structural material. Small cracks, up to $300\,\mu$m wide, are almost unavoidable in concrete and they accelerate its degradation by allowing the ingress of moisture. When concrete structures are subjected to bending they are reinforced with steel rods. Ingress of water into the concrete leads to corrosion of the steel. This is why concrete bridges have to be frequently inspected. Autonomous self-healing mechanisms can be introduced to heal cracks as large as $1\,$mm width, sometimes even larger. One such mechanism is based on the use of bacteria that produce calcium carbonate (limestone) at cracks and pores[15]. The limestone bonds well to cementitious products of various compositions. By filling cracks and pores it prevents ingress of water and it strengthens the concrete. The ureolytic bacteria strains can withstand the alkali environment (pH 13) of concrete. In the presence of water they decompose urea creating carbonate ions which react with calcium ions in the concrete to form calcium carbonate. However, calcium carbonate is brittle and if it is subjected to a varying load it will also crack. In such dynamic loading conditions polymer additions to the concrete that fill and heal cracks may be preferable.

9.6.2   Self-cleaning glass

Commercial self-cleaning glass has been available since 2001. The glass is coated by a thin film of titanium dioxide, TiO_2, typically less than $100\,$nm thick. When exposed to ultra violet (UV) radiation TiO_2 decomposes most organic compounds by a process called 'photocatalysis'. In addition, the TiO_2 becomes superhydrophilic on exposition to UV radiation. This enables water to spread over the coated glass forming an extremely thin water layer. As a result, rainwater washes off any residual contaminants on the coated glass and dries without leaving any water marks. It is the combination of the photocatalysis and superhydrophilicity of the TiO_2 coating that keeps the glass clean[16].

---

[15]de Belie, N, *et al.*, Advanced Materials Interfaces **5**, 1800074 (2018).

[16]Maiorov, V A, Glass Physics and Chemistry **45**, 161 (2019).

9.7   Closing remarks

In the preface to this book materials were defined as a subset of condensed matter having an application in some existing or intended technology. This chapter has examined the process by which materials are designed to meet the requirements of particular engineering applications. It has also touched on the use of self-assembly to fabricate materials that would otherwise be difficult to make, and how widely used existing materials have been enhanced by additions and surface treatments.

The design process begins with the *engineering* requirements of the material in its intended application. The creation of a material with the requisite properties uses the strong relations between properties, structure across the length scales, and fabrication process. That is where the *science* of materials comes to the fore. It does not necessarily entail the fabrication of a new material if a modification of an existing material meets the requirements, as in the example of self-cleaning glass.

Good design involves an understanding of the relevant materials science to meet the engineering requirements of the material. It reduces the time and cost of deploying a material in the intended application.

Materials discovery is somewhat different in that it involves finding materials with exceptional properties followed by a search for applications. Compared to materials design this is in reverse order. The science involved in materials discovery is data science and artificial intelligence. It is not clear how much materials science is involved, if any. Materials discovery is currently attracting a great deal of interest in universities and government research laboratories around the World. But in Europe, at least, there is considerable scepticism about it in industry.

Further reading

Ashby, Mike and Johnson, Kara *Materials and Design: The Art and Science of Material Selection in Product Design*, Elsevier (2014).

Ashby, M F, Shercliff, H and Cebon, D *Materials: Engineering, Science, Processing* *and Design*. 4th edition, Elsevier (2019).

Ball, Philip, *The self-made tapestry: Pattern formation in nature*, Oxford University Press (1999).

10
Metamaterials

Unusually for the physical sciences, negative refraction began life as a theoretical concept rather than an experimental discovery. It posed instead a challenge to experiment: to find materials with negative values of ϵ and μ. At the same time the theory had to be defended in debates on the validity of the concept. I think it is fair to say that 2003 saw this initial phase draw to a close with positive conclusions both on the concept and its experimental realisation. The future holds many new opportunities.

Pendry, J B, *Light and matter*, 36th Professor Harry Messel International Science School for High School Students, ed. C Stewart, Science Foundation for Physics of the University of Sydney (2011), p.134. Reprinted with permission.

10.1   Concept

Metamaterials are man-made materials created to manipulate waves of various kinds, from electromagnetic to sound waves. They are composite materials with carefully designed structural units much smaller than the wavelength of the waves they are designed to control but much larger than atoms. The properties of metamaterials are determined by their structures rather than their chemistry. The concepts of metamaterials have also been applied to the control of seismic waves and ocean waves.

10.2   Introduction

The prefix *meta* in metamaterial means 'beyond', suggesting that metamaterials are unlike other materials. In previous chapters we encountered a few of the very many man-made materials. They were carefully designed and fabricated, in some cases atom by atom. So in what sense are metamaterials unlike other man-made materials? Conventional materials derive their properties from the *chemical identity* of the atoms and molecules they are made of. Metamaterials derive their properties from strong coupling between *structural units* within them and an incident wave. These structural units are much larger than atoms but much smaller than the wavelength of the waves they interact with. Thus, the properties of metamaterials are determined by the carefully designed and assembled structural units within them rather than their chemical composition.

Metamaterials are also 'beyond' conventional materials in the sense that their unique properties often defy conventional wisdom. It was not thought possible to create any materials with some of their properties until metamaterials were developed. Metamaterials respond to waves in ways that have never been found with conventional materials. For example, elastic metamaterials have been created with effective negative masses and effective negative elastic moduli. Electromagnetic metamaterials have been created with effective negative refractive indices. No materials existed with these properties before metamaterials were developed around the turn of the millennium. The word 'effective' is significant. *In particular frequency ranges* the relevant properties – e.g. masses and elastic moduli for elastic waves and refractive indices for electromagnetic waves – are negative. At other frequencies they are positive, like conventional materials.

The key to understanding this dependence on frequency is the phenomenon of resonance. Resonance occurs when the frequency of a driving oscillating force coincides with the natural frequency of vibration of the system on which it acts. A familiar example is a mass m attached to a spring of stiffness k. Its natural frequency of vibration is $f_0 = (1/2\pi)\sqrt{k/m}$, which may be expressed in terms of the angular frequency $\omega_0 = 2\pi f_0$. If the mass is subjected to an oscillating force of constant amplitude and increasing angular frequency ω the amplitude of the oscillation of the mass is largest when $\omega = \omega_0$. When ω is less than ω_0 the mass moves in the same direction as the applied force: they are in phase with each other. But when ω is greater than ω_0 they are π out of phase: the mass is constantly moving in the opposite direction to the applied force. This aspect of resonant behaviour is at the heart of the design and properties of some of the most exciting metamaterials.

10.3   An example: a metamaterial for elastic waves

In this section we follow Milton and Willis[1] and consider how an elastic metamaterial responds to elastic waves. It comprises a rigid cylindrical rod of mass M_0 in which relatively small coaxial cylinders of height h have been carved out. The small cylinders are shaded pink in Fig. 10.1. At the centre of each cylinder there is a sphere of radius r and mass m attached by linear springs to the circular faces of its cylinder. Each spring has stiffness K, it has negligible mass and natural length $h/2 - r$. The total mass of the rod containing n such cylinders is $M_T = M_0 + nm$, where n is only 5 in Fig. 10.1 for clarity.

The rod is forced to undergo oscillations of angular frequency ω and amplitude A. Since the rod is rigid the forces experienced by each mass m are identical. The response of each mass m to the oscillating displacement of the rod is hidden from view by the surrounding rod. Only the response of the rod is measurable.

Let us select a small cylinder for detailed study. Let $X(t)$ and $x(t)$ be the positions of its left circular face and the centre of the mass m at time t. Then $X(t) = X_0 + A\sin(\omega t)$, where X_0 is the equilibrium position of the left circular face. The equilibrium position of the mass m is $X_0 + h/2$. The springs exert a force of $-2K(x(t) - X(t) - h/2)$ on the mass m. Applying Newton's second law to the mass m we have:

---

[1] Milton, G W and Willis, J R, Proc. R. Soc. A **463**, 855 (2007).

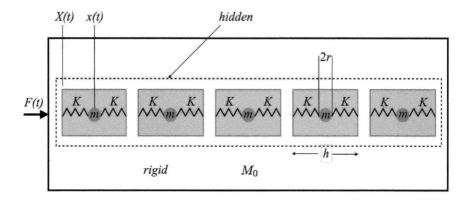

Fig. 10.1 A model of an elastic metamaterial. After Milton, G W and Willis, J R, Proc. R. Soc. A **463**, 855 (2007).

$$m\ddot{x} + 2K(x(t) - X(t) - h/2) = 0 \tag{10.1}$$

where $\ddot{x} = \mathrm{d}^2 x/\mathrm{d}t^2$ is the acceleration of the mass m and $X(t) = X_0 + A\sin(\omega t)$. This is the equation of a driven harmonic oscillator. In the absence of damping the solution for $x(t)$ must have the same oscillatory form as $X(t)$ with the same angular frequency ω: $x(t) = X_0 + h/2 + a\sin(\omega t)$. If the amplitude a of the oscillations is positive the mass m moves in the same direction as the rod, otherwise they move in opposite directions. It is determined by substituting this expression for $x(t)$ into eqn 10.1. We obtain:

$$\frac{a}{A} = \frac{1}{1 - \omega^2/\omega_0^2} \tag{10.2}$$

where $\omega_0 = \sqrt{2K/m}$ is the resonant angular frequency of the mass m. Thus at angular frequencies slightly above the resonant angular frequency a/A is large and *negative*: at any instant the masses m are then travelling in the *opposite* direction to the rod.

The momentum of the rod is $M_{eff}A\omega\cos(\omega t)$, where M_{eff} is the effective mass of the rod and the velocity of the rod is $\dot{X}(t) = A\omega\cos(\omega t)$. The (invisible) velocity of each mass m is $\dot{x}(t) = a\omega\cos(\omega t)$. Therefore,

$$M_{eff}A\omega\cos(\omega t) = M_0 A\omega\cos(\omega t) + nma\omega\cos(\omega t)$$

or

$$\frac{M_{eff}}{M_0} = 1 + \frac{nm/M_0}{1 - \omega^2/\omega_0^2}. \tag{10.3}$$

This relationship is sketched in Fig. 10.2. In the range of angular frequencies between zero and the resonant angular frequency the effective mass is greater than M_T. As the resonant frequency is approached from below the effective mass diverges to $+\infty$. The effective mass of the rod is zero when $\omega^2/\omega_0^2 = M_T/M_0$. At angular frequencies in the range $\omega_0^2 < \omega^2 < (M_T/M_0)\omega_0^2$ the effective mass is negative. In this range

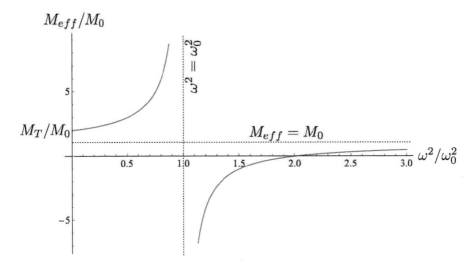

Fig. 10.2 Plot of eqn 10.3, for the case where $nm = M_0$, so that $M_T = 2M_0$. The asymptotes $\omega^2 = \omega_0^2$ and $M_{eff} = M_0$ are shown by broken lines.

the momentum and velocity of the rod are in opposite directions. As $\omega \to \infty$ we find $M_{eff} \to M_0$ because the masses m are unable to respond to such rapidly changing displacements of the rod. Notice also that when $\omega = 0$, the effective mass becomes M_T, because this is the rest mass of the rod. This limit is also reached when the spring stiffness $K \to \infty$ so that the masses m are held rigidly at the centre of each small cylinder, and $\omega_0 \to \infty$.

The assumption that the rod is rigid is a sensible approximation provided the wavelength of elastic waves in the matrix material is much greater than h. The model depicted in Fig. 10.1 ignores the damping of the masses m due to viscous drag. This may be modelled by including in the left hand side of eqn 10.1 a drag force $\eta \dot{x}(t)$, proportional to the velocity $\dot{x}(t)$. The constant drag coefficient η has dimensions of mass per unit time. The effective mass is then complex with real and imaginary parts:

$$\frac{M_{eff}}{M_0} = 1 + \frac{nm/M_0}{1 - \omega^2/\omega_0^2 - i\delta}, \tag{10.4}$$

where $\delta = \eta\omega/m\omega_0^2$ and $i = \sqrt{-1}$. This removes the unrealistic divergences to $\pm\infty$ of the real part of M_{eff} at the resonant angular frequency. Expressions analogous to eqn 10.4 appear in other types of metamaterials. The real and imaginary parts of M_{eff}/M_0 are plotted in Fig. 10.3.

This conception of an elastic metamaterial was introduced and realised experimentally by Sheng *et al.*[2]. They coated 1 cm balls of lead with 0.25 cm of silicone rubber and embedded them in an epoxy matrix. The balls of lead correspond to the masses m in the above model, the epoxy matrix corresponds to the rod and the silicon rubber

[2]Sheng, P, Zhang, X X,Liu, Z and Chan, C T, Physica B: Condensed Matter **338**, 201 (2003).

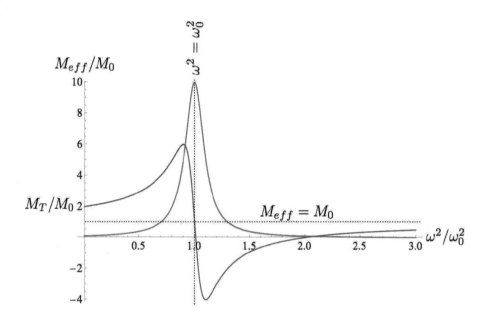

Fig. 10.3 The real (blue) and imaginary (red) parts of M_{eff}/M_0 in eqn 10.4, for the case where $nm = M_0$ and $\delta = 0.1$. The divergences to $\pm\infty$ in Fig. 10.2 at the resonant angular frequency have been removed in the real part of M_{eff}/M_0 by including viscous drag in eqn 10.1. Note there is still a range of frequencies where the real part of M_{eff}/M_0 is negative.

coating provides the viscoelastic coupling between the matrix and the balls with stiffness K and drag coefficient η. They observed strong coupling between elastic waves in the epoxy matrix and the localised vibrational motion of the lead balls, and the associated resonances.

10.4 Electromagnetic metamaterials and negative refraction

The discovery by James Clerk Maxwell in the 1860s that light is an electromagnetic wave is arguably the pinnacle of 19th century physics because it unified electricity, magnetism and optics. The electromagnetic spectrum spans from γ-rays, with a wavelength of order 10^{-12} m to radio waves, with a wavelength of 10^3 m or more. Visible light occupies a tiny range in this spectrum between 740 nm and 380 nm. An electromagnetic wave consists of electric and magnetic fields vibrating in phase at right angles to each other and to the direction of propagation of the wave. It is called a transverse wave because the vibrating fields are perpendicular to the direction of propagation, like ripples on water. In a longitudinal wave the vibrating field is parallel to the direction of propagation. An example of a longitudinal wave is a sound wave in air which comprises regions of compression and expansion of the air along the direction of its propagation.

Maxwell showed that the speed of light, $c_0 = 2.998 \times 10^8$ ms^{-1}, in a vacuum is equal to $1/\sqrt{\epsilon_0\mu_0}$, where $\epsilon_0 = 8.854 \times 10^{-12}$ CV^{-1}m^{-1} and $\mu_0 = 4\pi \times 10^{-7}$ VC^{-1}s^{2}m^{-1} are

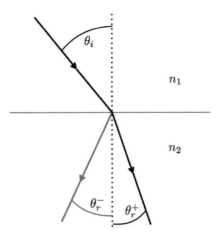

Fig. 10.4 To illustrate Snell's law of refraction. A ray of light inclined to the vertical at an angle θ_i in a medium of refractive index n_1 is incident on an interface (green) where it enters a medium of refractive index n_2. If $n_2 > 0$ the ray is refracted at an angle θ_r^+ to the vertical. If $n_2 < 0$ it is refracted (red) at an angle θ_r^- to the vertical.

the permittivity and permeability of free space respectively[3]. When light enters matter its velocity[4] is reduced to c_0/n where n is the refractive index. The refractive index varies with the frequency of light, which is why rain drops can separate sunlight into its constituent colours in a rainbow.

Refraction of light is a familiar phenomenon. When we look into a swimming pool it looks shallower than it is. Snell's law describes the relationship between the angle of incidence, θ_i, and the angle of refraction, θ_r, when light passes from an isotropic medium with refractive index n_1 through an interface into another isotropic medium with refractive index n_2:

$$\frac{\sin \theta_i}{\sin \theta_r} = \frac{n_2}{n_1}. \tag{10.5}$$

This is illustrated in Fig. 10.4. Snell's law follows from the translational symmetry of the interface. We learnt in Chapter 5 that continuous translational symmetry leads to conservation of momentum. Therefore, the momentum of the light wave parallel to the interface is conserved. The momentum of the light wave is given by the de Broglie relationship h/λ, where λ is its wavelength, and it is directed along the ray. The frequency f of the light wave does not change on crossing the interface. From the definition of the refractive index the speeds of light in media 1 and 2 are c_0/n_1 and c_0/n_2 respectively. Therefore the wavelengths are $\lambda_1 = c_0/(n_1 f)$ and $\lambda_2 = c_0/(n_2 f)$. The components of the momentum parallel to the interface are $(h/\lambda_1) \sin \theta_i = (h n_1 f / c_0) \sin \theta_i$ and

---

[3]The quoted values of ϵ_0 and μ_0 are in SI units, but I have chosen to express them in terms of C = Coulomb (unit of charge), V= Volt, m = metre and s=second, because then it is obvious that the product $\epsilon_0 \mu_0$ has the dimensions of inverse square velocity. Note that the SI unit of capacitance is the Farad (F) which is one Coulomb per Volt.

[4]More precisely this is the phase velocity.

$(h/\lambda_2) \sin \theta_r = (h n_2 f / c_0) \sin \theta_r$. By equating these parallel momenta we obtain Snell's law.

The refractive index is a property of the material and it depends on the angular frequency ω of the incident electromagnetic wave. It is given by $n^2(\omega) = \epsilon_r(\omega) \mu_r(\omega)$. In this equation ϵ_r is[5] the ratio of the electric permittivity of the material to ϵ_0. Similarly, μ_r is the ratio of the magnetic permeability of the material to μ_0. The speed of the electromagnetic wave in the material is $c_0/n = 1/\sqrt{\epsilon_r \epsilon_0 \mu_r \mu_0}$. The degree to which the material can be polarised electrically and magnetically, at the angular frequency of the incident electromagentic wave, determines $\epsilon_r(\omega)$ and $\mu_r(\omega)$ respectively.

The story of electromagnetic metamaterials begins with a paper by Victor Veselago in 1967[6]. He posed a seminal question: what would happen to the propagation of electromagnetic waves in a material if both ϵ_r and μ_r were negative? He showed theoretically that the refractive index would also be negative. Light entering such a material would be refracted along the red ray shown in Fig. 10.4. We shall see below that achieving $\epsilon_r < 0$ in practice over a limited range of frequencies is not difficult. But it was not until the development of metamaterials around the turn of the millennium that materials with $\mu_r < 0$ were created.

The shiny appearance of metals stems from the inability of light to propagate through them, as in a mirror where a thin layer of aluminium reflects most of the light incident upon it. The optical properties of most metals are dominated by the response of their free electrons to incident electromagnetic waves. Free electrons experience collectively a time-dependent force $-eE(t)$ from the oscillating electric field $E(t) = E_0(\omega)e^{-i\omega t}$ of the electromagnetic wave, where $E_0(\omega)$ is the amplitude of the oscillating electric field of angular frequency ω and $-e$ is the electron charge. If we ignore the viscous damping of the free electron response due to collisions, Newton's second law for each free electron in this collective motion is as follows:

$$m\ddot{x} = -eE_0(\omega)e^{-i\omega t}.$$

Here m is the electron mass and x is the collective displacement of the electrons relative to the immobile positive background charge of the metal ion cores. The solution to this equation is readily obtained by substituting $x = x_0(\omega)e^{-i\omega t}$. We obtain:

$$x(t) = \frac{e}{m\omega^2} E(t). \tag{10.6}$$

This collective displacement sets up a wave of electric polarisation per unit volume:

$$P(t) = P_0(\omega)e^{-i\omega t} = -n_e e x(t) = -n_e e x_0(\omega)e^{-i\omega t} = -\frac{n_e e^2}{m\omega^2} E_0(\omega)e^{-i\omega t},$$

where n_e is the number of free electrons per unit volume. The dielectric function $\epsilon_r(\omega)$ is defined by $P_0(\omega) = (\epsilon_r(\omega) - 1)\epsilon_0 E_0(\omega)$. It follows that:

$$\epsilon_r(\omega) = 1 - \frac{\omega_p^2}{\omega^2}, \tag{10.7}$$

---

[5] It is also called the dielectric function of the material.
[6] English translation: Veselago, V G, Soviet Physics Uspekhi **10**, 509 (1968).

where $\omega_p = \sqrt{ne^2/(m\epsilon_0)}$ is called the volume plasma frequency of the free electron gas. In aluminium $\omega_p \approx 2.5 \times 10^{16}$ Hz. The range of angular frequencies of visible light is approximately from 2.5×10^{15} to 5×10^{15} Hz. It follows that $\epsilon_r(\omega) < 0$ for visible light, and it becomes positive only in the ultraviolet. This is the reason why metals reflect rather than transmit visible light. As we noted above this simple analysis ignores electron energy losses creating resistance. When it is included it is found that $\epsilon_r(\omega)$ of naturally occurring metals is dominated by energy losses at frequencies in the infrared and below.

If ω_p could somehow be reduced to ≈ 1 GHz $= 10^9$ Hz the material would become transparent to electromagnetic waves above these angular frequencies. Below these frequencies ϵ_r would be negative. A metamaterial structure to achieve this goal was proposed theoretically by Pendry *et al.*[7] using three dimensional arrays of continuous, thin metallic wires embedded in a non-conducting matrix. The dramatic reduction of ω_p was achieved by (a) the reduction of the free electron density n by confining the free electrons to thin wires separated by distances much larger than their diameter, and (b) the increase of the effective mass m of the electrons through the self-inductance of the wires which opposed changes in the induced electric currents in the wires. This design was subsequently confirmed experimentally[8] using arrays of 20 μm diameter gold plated tungsten wires spaced 5 mm apart in a polystyrene matrix. The plasma frequency was found to be around 9 GHz as predicted. The wire spacing was approximately ten times less than the wavelength of the electromagnetic wave with which the metamaterial was designed to interact. Consequently, the incident electromagnetic waves were not diffracted by the wires. This metamaterial illustrates the point that its dielectric function is determined by its carefully designed structure rather than its chemistry.

Designs of metamaterials to manipulate the magnetic permeability $\mu_r(\omega)$ were proposed in another paper by Pendry *et al.*[9] using 'split ring resonators'. The goal was to develop a metamaterial with a negative $\mu_r(\omega)$ response to electromagnetic waves of frequencies in the microwave range where the wavelength is of order 1 cm. Furthermore they wanted the magnetic response to be the same in different directions within the material. They achieved these goals by embedding a lattice of split ring resonators made of thin sheets of nonmagnetic metal in an inert matrix. The electric currents induced in these structures by the oscillating magnetic fields of the incident microwaves created additional magnetic fields that enabled the material to respond with a different (effective) magnetic permeability. The key to the design was to create a structure with a resonant magnetic response at a frequency of microwave radiation. The split ring resonator has both an inductance and a capacitance and hence a resonant frequency where the magnetic response is enhanced considerably. At frequencies just below the resonance $\mu_r(\omega)$ is positive. But at frequencies just above the resonance $\mu_r(\omega)$ is negative. There are no naturally occurring materials with $\mu_r(\omega) < 0$ at these

---

[7] Pendry, J B, Holden, A J, Stewart, W J and Youngs, I, Phys. Rev. Lett. **76**, 4773 (1996).

[8] Pendry, J B, Holden, A J, Robbins, D J and Stewart, W J, J. Phys.: Condens. Matter **10**, 4785 (1998).

[9] Pendry, J B, Holden, A J, Robbins, D J and Stewart, W J, IEEE Transactions on Microwave Theory and Techniques **47**, 2075 (1999).

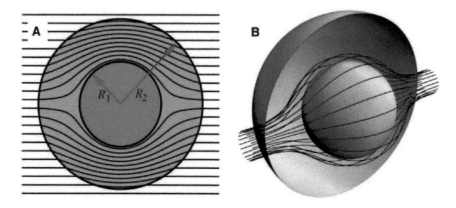

Fig. 10.5 (A) Two dimensional cross-section of light rays deflected around a sphere of radius R_1 by a metamaterial cloak occupying the shell between R_1 and R_2. (B) A three-dimensional view of the same process. An object inside the inner sphere is invisible to an observer outside the outer sphere. From Pendry *et al.*[11]. Reprinted with permission from AAAS.

frequencies. For the first time, the development of metamaterials held out the possibility of creating materials with both $\epsilon_r(\omega) < 0$ and $\mu_r(\omega) < 0$, and hence a negative refractive index in the range of frequencies where both inequalities are satisfied.

In 2000 Smith *et al.*[10] published the first experimental realisation of a metamaterial with a negative $\mu_r(\omega)$ and a negative $\epsilon_r(\omega)$ at microwave frequencies. Their design was based on Pendry's ideas of using split ring resonators to achieve negative $\mu_r(\omega)$ and arrays of continuous thin wires to achieve negative $\epsilon_r(\omega)$. They had created the first material with a negative refractive index, albeit in a limited range of microwave frequencies. Interest in metamaterials exploded.

10.5   Invisibility cloaks

Progress in research on electromagnetic metamaterials has been rapid. By 2006 Pendry *et al.*[11] were able to claim it was conceivable that a metamaterial could be designed and constructed with independent and arbitrary values of permittivity and permeability throughout. To illustrate what could be achieved with electromagnetic metamaterials they decided to design theoretically[11] and manufacture[12] an 'invisibility cloak'. This certainly captured the attention of the World media and the public imagination at a time when fans of Harry Potter[13] were admiring his fictional cloak of invisibility.

The goal is to conceal an object in such a way that observers are unaware that something has been hidden from them. A metamaterial cloak guides light rays around

---

[10]Smith, D R, Padilla, W J, Vier, D C, Nemat-Nasser, S C and Schultz, S, Phys. Rev. Lett. **84**, 4184 (2000).

[11] Pendry, J B, Schurig, D and Smith, D R, Science **312**, 1780 (2006).

[12]Schurig, D, Mock, J J, Justice, B J, Cummer, S A, Pendry, J B, Starr, A F and Smith, D R, Science **314**, 977 (2006).

[13]Harry Potter is the principal character in a popular series of novels by J K Rowling and the subsequent films.

the object so that they return to travelling in the same direction, as illustrated in Fig. 10.5. To an external observer this inner region is empty because light is guided around it, but an object may be concealed within it. The cloak does not reflect incident waves and it does not cast a shadow, either of which would reveal its presence.

With the ability to change the local ϵ_r and μ_r of the cloak at will it is clear that a metamaterial could in principle be created to guide electromagnetic waves of the relevant frequencies around the object to be concealed. The question is how to determine the spatial variations of $\epsilon_r(x, y, z)$ and $\mu_r(x, y, z)$ required to achieve this. Pendry *et al.*[11] solved this problem by asking the question in a different way. Suppose we deform space by a coordinate transformation such that the space occupied by the inner sphere $(r < R_1)$ in Fig. 10.5 is squeezed into the shell between $r = R_1$ and $r = R_2$ which is the region to be occupied by the cloak. By deforming space we might expect the fundamental (Maxwell) equations of electromagnetism to change beyond all recognition. It turns out they retain the same form but the dielectric function and magnetic permeability become spatially dependent and anisotropic[14] in the region $(R_1 < r < R_2)$ occupied by the cloak. We can use those spatially varying dielectric functions and magnetic permeabilities to achieve the same goal of guiding light around the inner sphere. This trick of deforming space is called 'transformation optics'.

The experimental realisation[12] of a two-dimensional invisibility cloak for microwaves used a design based on transformation optics. It involved the use of split ring resonators in a pattern that was not cubic nor even periodic. The experiment proved the principle.

10.6 Closing remarks

The development of metamaterials has extended the concept of a material. Their responses to waves of various kinds depend on their structure rather than their chemical composition. The structural elements of metamaterials are not atoms or molecules but entities much larger than atoms and yet smaller than the wavelength of the radiation they are intended to control. To control visible light these entities have to be at the nanoscale and their fabrication involves nanotechnology. At very much larger length-scales the ideas of metamaterials have been introduced into the protection of buildings and even cities from earthquakes by using carefully designed arrays of empty boreholes and boreholes filled with resonant inclusions to dampen seismic waves[15]. Similar ideas are being developed to protect coastal communities from tsunami ocean waves. This is another illustration of the cross-fertilisation of ideas between materials science and earth sciences. The field of metamaterials is still in a state of rapid development and it is likely there will be further surprises and applications.

Metamaterials were conceived by original, ground-breaking thinking within the laws of physics. They are paragons of materials design. Transformation optics was created to design materials with the desired properties, such as an invisibility cloak. Could data science and artificial intelligence have discovered metamaterials? I cannot see how. It took the imagination and real intelligence of John Pendry and David Smith to conceive, design and fabricate metamaterials.

[14]Ward, A J and Pendry, J B, J. Mod. Opt. **43**, 773 (1996).

[15]Miniaci, M, Krushynska, A, Bosia, F and Pugno N M, New J. Phys. **18**, 083041 (2016).

Further reading

Cai, W and Shalaev, V, *Optical metamaterials*, Springer (2010).

Pendry, J B, in *Light and matter*, 36th Professor Harry Messel International Science School for High School Students, ed. C Stewart, Science Foundation for Physics of the University of Sydney (2011).

11
Biological matter as a material

What I cannot create I do not understand.

Richard Feynman, written on the blackboard in his office and found soon after his death.

11.1 Concept

Viewed as a material, biological matter displays self-organisation through the collective action of energy-consuming agents. The creation of living matter from non-living matter is an ultimate goal of science.

11.2 What is life?

In Chapter 3 we saw that diffusion in solids and liquids is the result of chaotic motion of atoms. But at a larger length scale diffusion is orderly and deterministic – it obeys a differential equation. This is an example of the success of statistical physics where a law of physics at a larger length scale emerges from the chaotic behaviour of atoms at a much smaller length scale. Another example would be the pressure exerted by a gas in a container which arises from the exchange of momentum when gas atoms bounce off its walls. In his book[1] Schrödinger contrasted the approach of statistical physics with the unique genetic information stored in a cell that controls its function and contains hereditary information. It was this contrast that led Schrödinger to conclude that the physics of living matter was quite different from the statistical physics of large groups of atoms and molecules that had been so successful up to then. By coincidence his book was published in the same year as the discovery[2] that DNA is the molecule in which genetic information is stored.

A number of pioneers of molecular biology have acknowledged the inspiration of Schrödinger's book. They include Francis Crick and James Watson who, together with Maurice Wilkins, were awarded the Nobel Prize in Physiology or Medicine in 1962 for unravelling the structure of DNA. But not everyone was impressed by it. In 1987 the molecular biologist Max Perutz, who was awarded the Nobel Prize for Chemistry in

---

[1] Schrödinger, Erwin *What is Life?*, Cambridge University Press (1944).
[2] Avery, O T, MacLeod, C M and McCarty, M, Journal of Experimental Medicine **79**, 137 (1944).

1962, gave the following assessment[3]:

Sadly, however, a close study of his book and of the related literature has shown me that what was true in his book was not original, and most of what was original was known not to be true even when the book was written. Moreover, the book ignores some crucial discoveries that were published before it went into print ... The apparent contradictions between life and the statistical laws of physics can be resolved by invoking a science largely ignored by Schrödinger. That science is chemistry.

We can usually recognise the difference between animate and inanimate matter very easily. Biologists have identified seven things that all living things do: they respond to stimuli, grow over time, reproduce themselves, they regulate their body temperature, metabolise food, comprise one or more cells and adapt to their environment. However, the identification of what living things do is not the same as defining what life is, and it also does not explain how life emerged. It is surprisingly difficult to define life. For example, we might say the ability to reproduce is the essential quality of life. But a mule cannot reproduce itself because it is infertile. On the other hand a virus can replicate itself but only with the machinery and metabolism of a host cell. Is a virus alive or not? It appears the jury is still out. The search for extraterrestrial life by NASA needs at least a working definition of life because without it how will NASA know whether they have found it? NASA's working definition of life is: 'Life is a self-sustaining chemical system capable of Darwinian evolution.' As a working definition it may be limited in its usefulness because Darwinian evolution may take more time than is available to make a judgement. In his recent book[4] the Nobel prize winner Paul Nurse introduces five big ideas of biology from which he draws out three principles to define life:

1. Living organisms must be able to evolve through natural selection. This requires them to have the ability to reproduce, they must have a hereditary system, and their hereditary system must be capable of variability.
2. Life forms are contained and separate from, but in communication with, their environments, like a cell.
3. Life forms are chemical and physical machines that receive and respond to information in a purposeful manner. The purpose may be to protect themselves, to reproduce, to find food and so on.

There is now a resurgence of interest in the 'physics of life', with the creation of new research networks and institutes. In an editorial in 2019 in *Physical Review Let-*

---

[3]Perutz, M F, in *Schrödinger: Centenary Celebration of a Polymath*, edited by Kilmister, C W, pages 234-251, Copyright © Cambridge University Press 1987. Reproduced with permission through PLSclear.

[4]Nurse, Paul *What is Life?* David Fickling Books (2020).

ters Grill and Chaté wrote[5]:

> ...*we must go further and encompass the* <u>material science</u> *of animate matter, the complex processes of self-organization, as well as information-theoretical aspects.* (my underlining)

As part of this resurgence there is a growing recognition of the role of collective phenomena[6] in biology arising from interactions between smaller components far from thermodynamic equilibrium, giving rise to the complexity evident in biological systems. This represents a change of direction from the reductionist approach of the very successful fields of molecular biology and single-molecule biophysics to a more holistic view of organisms as complex systems. It has similarities to the discussion in Chapter 8 of the role of the collective behaviour of defects in materials driven far from equilibrium and the emergence of new phenomena and new materials physics. To view life only as an enormous set of chemical reactions at the molecular level may be as limiting as the view that the properties of non-living materials may be understood by studying them only at the atomic scale. Systems biologists view living organisms as complex systems that process information about themselves and their environments through the myriad chemical processes taking place within each cell. Detailing the complexity is not an end in itself. It has to explain how a cell responds to information to achieve some purpose, to provide meaning to the complexity.

11.3 Active matter

Biological cells are dynamic objects. They change shape, divide and move. The cytoplasm is the material between the nucleus and the cell membrane and it is around 80% water. The shape and rigidity of the cell is governed by a network of filaments made of certain protein molecules in the cytoplasm. The network behaves like a scaffold supporting the cell and it is called the cytoskeleton. The cytoskeleton determines how the cell responds mechanically to applied forces. The filaments of the cytoskeleton are also conduits along which motor proteins convey cargos of other proteins and cell structures[7]. Filaments of the cytoskeleton are polymers consisting of repeating sequences of protein molecules. By assembling and disassembling filaments through polymerisation and depolymerisation reactions the cell can change shape and move. It can also exert forces on neighbouring cells.

Treating the cell as a material, it is interesting to investigate its mechanical properties. A variety of experimental techniques have been used to do this[8]. It is found that cells display nonlinear elastic behaviour and their continued deformation under

---

[5] Reprinted excerpt with permission from Grill, S W and Chaté, H, Physical Review Letters **123**, 130001 (2019). Copyright (2019) by the American Physical Society.

[6] Goldenfeld, N and Woese, C, Annual Review of Condensed Matter Physics **2**, 375 (2011).

[7] Motor proteins get the energy they need to do this from the chemical energy released when ATP, (adenosine triphosphate) is hydrated to make ADP (adenosine diphosphate) or AMP (adenosine monophosphate).

[8] Kasza, K E, Rowat, A C, Liu, J, Angelini, T E, Brangwynne, C P, Koenderink, G H and Weitz, D A, Current opinion in cell biology **19**, 101 (2007).

a constant force indicates they also display viscous behaviour. Kinks are introduced into some of the more flexible polymer filaments through thermal fluctuations. As these kinks are pulled straight under the action of an applied force the elastic modulus increases. At time scales of more than 30 seconds the cytoskeleton restructures and this gives rise to additional relaxation and viscous behaviour. The ability of the cytoskeleton to adapt through new polymerisation and depolymerisation reactions at these longer time-scales enables the cell to respond to forces in a manner unlike any other polymer network.

In contrast to non-living matter, living matter is continuously consuming energy from its environment to power its various functions. Molecules self-assemble to create machines inside a cell. The cytoskeleton structures and restructures the cell interior enabling the cell to change shape and move. Cells come together to form tissues. Tissues self-organise to create organs and organisms. At each hierarchical level of organisation new biological functions emerge from the collective action of agents at smaller scales operating away from thermodynamic equilibrium. The field of active matter, 'at the interface between materials science and cell biology'[9], attempts to understand how the collective action of self-driven, energy-consuming agents leads to the self-organisation in cells and larger biological structures. There are obvious similarities between active matter and the emergence of self-organised defect structures in crystals driven far from equilibrium, such as in work-hardening (see Chapter 8). One essential difference is that the self-organising agents in biology are self-driven machines consuming energy supplied through their environment. Another is that the agents acting in biology are purposeful.

To study the processes of self-organisation in the cytoskeleton model active matter systems have been synthesised[10] consisting only of the proteins making up the biopolymers and motor proteins in solutions containing ATP and GTP[11]. ATP fuels the molecular motors and the polymerisation reactions. GTP fuels the depolymerisation reactions. The biopolymers have an intrinsic electric polarisation which forces the molecular motors to migrate along them only in one direction, some towards the positive end, others the negative end. Myosin motors move towards the positive end of actin polymers, which are called microfilaments. Kinesin and dynein motors walk along tubulin polymers, which are called microtubules. Motors may be either processive or non-processive. Processive motors may make many steps before detaching from the polymer and a single motor molecule can transport a cargo over a large distance. Kinesin and dynein motors are processive, with kinesin motors moving to the positive end and dynein motors moving to the negative end of a microtubule. Exceptionally the kinesin-14A family of motors, of which Ncd is a member, travels towards the negative end of a microtubule. Most processive motors are attached to the polymer by two molecular 'heads' and walk along the polymer by the heads stepping past one another, in a 'hand-over-hand' way, alternating the lead position. Non-processive motors leave the polymer after one step. When there are large numbers of them they can still move cargos over significant distances. Motor proteins reorganise and restructure

[9]Needleman, D and Dogic, Z, Nature Reviews Materials **2**, 17048 (2017).

[10]Surrey, T, Nédélec, F, Leibler, S and Karsenti, E, Science **292**, 1167 (2001).

[11]GTP is guanosine triphosphate

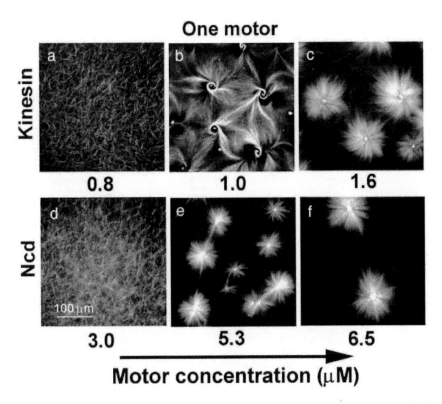

Fig. 11.1 Dark field optical micrographs of self-organised structures of microtubules produced in solutions of tubulin protein (the monomer constituent of microtubules) and *either* kinesin (a-c) *or* Ncd (d-f) motor proteins of varying concentrations, together with ATP and GTP to fuel the process. From Surrey, T, Nédélec, F, Leibler, S and Karsenti, E, Science **292**, 1167 (2001). Reprinted with permission from AAAS.

the biopolymer network that forms the cytoskeleton by creating cross-links between polymer molecules enabling them to slide relative to each other.

Fig. 11.1 shows the structures produced in vitro by mixing tubulin proteins (the monomers of microtubules) and one type of motor protein, either kinesin (a-c) or Ncd (d-f) as a function of the concentration of motor protein. At the lowest concentration of kinesin molecular motors, Fig. 11.1a, the microtubules are randomly arranged. At slightly higher concentration (b) of kinesin vortices of microtubules form. At even higher concentration (c) star-like structures of microtubules appear, called asters. In contrast only asters are formed with Ncd motors (e) and (f). The asters in the kinesin and Ncd organised structures have different orientations of the microtubules. The negative (positive) ends of the microtubules pointed towards the centre of the asters produced with Ncd (kinesin) motors. In both cases the motor proteins collected at the centres of the asters.

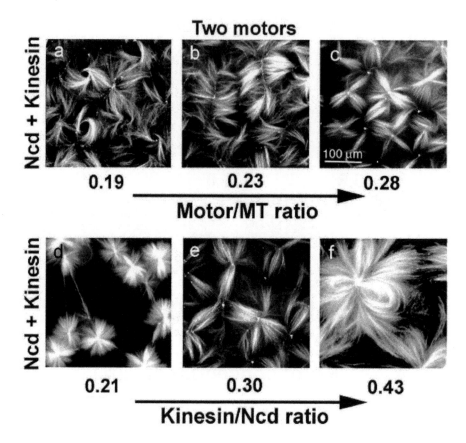

Fig. 11.2 Dark field optical micrographs of self-organised structures of microtubules produced in solutions of tubulin protein by *both* kinesin *and* Ncd motor proteins, together with ATP and GTP to fuel the process. a-c: variation of the motor/tubulin ratio at constant Ncd/kinesin concentration ratio. The concentrations of kinesin, Ncd, and tubulin were (a) 1.2, 4.0, and 28 μM; (b) 1.5, 4.9, and 28 μM; (c) 1.7, 5.6, and 26 μM. d to f: variation of the kinesin/Ncd concentration ratio. Kinesin and Ncd concentrations were (d) 1.2 and 5.6 μM; (e) 1.7 and 5.6 μM; (f) 2.0 and 4.6 μM. The tubulin concentration was 28 μM. From Surrey, T, Nédélec, F, Leibler, S and Karsenti, E, Science **292**, 1167 (2001). Reprinted with permission from AAAS.

Fig. 11.2 shows the self-organised microtubule structures produced by the simultaneous action of kinesin and Ncd protein motors in a solution of tubulin proteins and ACT and GCT. These motors move towards opposite ends of the microtubules. At the lowest ratio of concentrations of motors to tubulin proteins the kinesin motors produced vortices and the Ncd motors produced asters, see Fig. 11.2a. This is consistent with Figs. 11.1b and e. In Figs. 11.2b and c we see the effect of increasing the ratio of the total motor concentration to the tubulin concentration, while maintaining the ratio of Ncd to kinesin concentrations constant. A network of 'poles' is produced, resembling hair partings, which eventually leads at higher ratios of motor to tubulin concentrations to asters of either Ncd or kinesin. Figs. 11.2d-f show the effect of increasing the ratio of kinesin to Ncd motor proteins at a constant concentration of tubulin.

These experiments have demonstrated that self-organisation of biological structures by energy-consuming agents can be studied outside cells. They have stimulated new theoretical and computational studies, and opened a new area of materials science.

11.4   Synthetic biology

Some scientists believe we will understand what life is and how it came about only when we are able to create it from inanimate matter – a belief that chimes with the quote by Feynman at the beginning of this chapter.

A first step along this path was taken in 2010 by Craig Venter and colleagues[12] who created the first living cell controlled by synthetic DNA. They reconstructed the genome of a common bacterium called Mycoplasma mycoides by first producing molecular strands of the bacterium's DNA. These strands were considered the minimum necessary for life. The strands were reassembled into a whole by inserting them into yeast followed by E coli bacteria. The innate repair mechanism of these bugs treated the strands as broken fragments and assembled them into a whole genome. To label the assembled genome as synthetic the researchers inserted some sequences of coded messages into the reassembled genome, including the quote by Feynman at the beginning of this chapter. These 'watermark' sequences did not affect the functioning of the final organism, but their presence in its progeny indicated that the synthetic genome was being passed on. The synthetic genome was transplanted into another bacterium, Mycoplasma capricolum, that had had its genome removed. As soon as the new genome was transplanted it was 'read' by the cell to make a new set of proteins, and in a short time all the characteristics of the Mycoplasma capricolum disappeared and those of Mycoplasma mycoides emerged. In this way the researchers recreated an existing bacterial life-form, Mycoplasma mycoides, but with a man-made genome. All the machinery of the synthetic cell, other than its genome, came from a pre-existing cell of Mycoplasma capricolum. This was a remarkable advance but it is still a long way from creating new life.

The machinery of a cell is extremely complex, as described lucidly in Nurse's book. There are efforts to create artificial cells, most recently with a 'bottom up' approach[13] using components that are well understood. But so far no completely artificial cell has been made that satisfies Nurse's three principles defining life.

11.5   Closing remarks

The variety of self-organised structures that can emerge from the collective action of energy-consuming biological agents is truly remarkable. Both biological and non-biological matter display the emergence of new structures at larger length scales from

[12]Gibson, D G, Glass, J I, Lartigue, C *et al.*, Science **329**, 52 (2010).
[13]Powell, K, Nature **563**, 172 (2018).

the collective action of agents at smaller length scales. In biological matter the new structures have purpose and meaning.

The creation of life, as defined by Nurse's three principles, from inanimate materials remains one of the greatest challenges in science.

Further reading

Davies, J A, *Synthetic biology: A very short introduction.* Oxford University Press (2018).

Davies, Paul *The demon in the machine: How hidden webs of information are solving the mystery of life*, Allen Lane (2019).

Nurse, Paul *What is Life? Understand biology in five steps*, David Fickling Books (2020).

Index

absolute zero, 3
activation energy, 32
active matter, 127
adatom, 41
albite, 56
angular momentum
  orbital, 87
  quenching, 87
artificial intelligence, 103, 113, 123
atomic motion, 3
atoms
  size and identity, 65

band gaps, 74
band theory, 74
  classification of metals and insulators, 74
  failures, 74
  non-crystalline materials, 74
  semiconductors, 74
Bohr model of hydrogen, 66
Boltzmann constant, 9
bosons, 71
broken symmetry, 53
Brownian motion, 33
  and diffusion, 33
bubble raft, 106
Burgers circuit, 48
Burgers vector, 47

catalysis, 86
catalyst
  selectivity, 86
cementite, 57
chemical potential, 4, 15
chemical work, 16
collective behaviour
  in biological matter, 127
  in inanimate materials, 94
colloids, 9
complex system, 104, 127
conduction electrons, 73
conservation laws, 54
core electrons, 73
core-shell particles, 107
crystal momentum, 55
crystal symmetries
  rotational, 52
  translational, 52
Curie temperature, 88
cytoplasm, 127

cytoskeleton, 127

data science, 103, 113, 123
de Broglie relation, 55
defects
  agents of change, 40, 41
  cracks, 99
  disclinations, 57
  dislocations, 46, 57
  grain boundaries, 50
  point, 41
  topological, 57
deformation
  elastic, 46
  elastic limit, 46
  plastic, 46
density of electronic states, 89
diffusion, 33
  activation free energy, 42
  along grain boundaries, 51
  and equilibrium fluctuations, 33
  and mobility, 35
  and oxidation, 40
  as a random walk, 33
  as Brownian motion, 33
  at thermodynamic equilibrium, 33
  by direct exchange, 43
  by ring mechanism, 43
  by self-interstitials, 42
  by vacancy mechanism, 41, 43
  distance in time t, 34
  driving force, 34
  in ionic crystals, 45
  in sintering, 41
  mechanisms, 42
  predominance of vacancy mechanism, 43
  quantum, 79
  reptation, 37
  surface, 41
diffusivity, 34
disclination, 57
  wedge, 58
dislocation, 46
  agent of plastic deformation, 50
  as a topological defect, 57
  Burgers circuit, 48, 58
  Burgers vector, 47, 58
  core structures, 49
  cut, 57
  edge, 47

in a quasicrystal, 63
in earth sciences, 50
kinks, 48
line bounding slipped region, 47
localisation of slip, 47
loop, 47
mixed, 48
nucleation of cracks, 50
pileup, 50
screw, 48
self-organisation, 98
shielding, 100
work hardening, 98
domain, 54, 89
double slit experiment, 66
    with C_{60} molecules, 69
    with electrons, 68
drift speed, 36
driven materials, 2, 104
Dulong-Petit law, 78
dynamic equilibrium, 31

edge dislocation, 47
    as a wedge disclination dipole, 58
Einstein relation, 35
elastic constants, 56
electric permittivity, 120
electrical resistance in a metal, 36
electromagnetic spectrum, 118
electromagnetic wave, 118
electron microscope, 70
electron screening in metals, 75
electron-hole pair, 84
electronic conduction, 96
    ballistic transport, 96
    conductance quantisation, 96
    diffusive transport, 96
    mean free path, 96
    multiscale nature, 96
    Ohm's law, 96
    scattering, 96
electrostrictive effect, 112
emergence, 95
    in biological organisms, 128
energy, 4
    conservation, 4
    degradation, 8
    dispersal and entropy, 8
    forms, 4
    Gibbs free energy, 14
    heat, 3
    Helmholtz free energy, 13
    internal, 4
    potential, 3
    work, 4
entropy, 5, 6
    and energy dispersal, 8
    and heat transfer, 7
    and number of microstates, 8
    Boltzmann equation, 9

configurational, 10
direction of time, 5, 12
function of state, 7
ideal gas, 9
increase in a closed system, 12
maximum, 6
microstates, 8
of a solid, 9
stabilises open crystal structures, 10
exchange interaction, 72
    magnetic moment of free atoms, 87
excited state
    probability, 32
exciton, 84
exclusion principle, 71
    consequences, 72
    Hund's rule, 87

Fermi energy, 75
Fermi wavelength, 75
fermions, 72
ferromagnetism, 87
    in iron, 88
fluctuation-dissipation theorem, 34
fluctuations, 31
    $1/\sqrt{N}$ dependence, 32
    and diffusion, 33
    distributions of nanoparticle sizes, 110
    of voltage, 36
fluorescence, 85
fracture, 99
    brittle, 99
    brittle to ductile transition, 100
    ductile, 100
    grain boundary embrittlement, 100
    Griffith's criterion, 99
    multi-scale nature, 100
    shielding dislocations, 100
    with plasticity, 99
Frank-Kasper phases, 60
free electron approximation, 74
free energy
    against composition, 21
    Gibbs, 14
    Helmholtz, 13
    ideal solution, 22
    mixing favourable, 22
    phase separation, 22
    random alloy, 22
Frenkel defect, 45

giant magnetoresistance (GMR), 90
Gibbs free energy, 14
    and chemical potentials, 16
    and phase equilibrium, 15
    and thermal activation, 32
Gibbs phase rule, 17
Gibbs-Duhem equation, 16
grain boundary, 41, 50, 53, 58, 77
    agent of recrystallisation, 51

barrier to slip, 50
channel for rapid diffusion, 51
cohesion enhancers, 100
compatibility stresses, 51
embrittlement, 100
heterogeneous nculeation site, 51
slip transmission, 50
grain growth, 51
grain refinement, 51
grains, 50
graphene, 57
gyromagnetic ratio, 87

halochromic effect, 111
heat, 30
and conservation of energy, 30
as caloric, 30
as energy, 30
as kinetic energy of atoms, 30
equivalence to work, 30
heat reservoir, 12
Helmholtz free energy, 13
hole
a missing electron, 84
Hund's rule, 87
hydrogen atom
quantum numbers, 72
hydrogen embrittlement, 79

icosahedral symmetry, 60
ideal gas, 3
incommensurate waves, 62
independent elastic constants
in albite, 56
in cementite, 57
in cubic crystals, 57
in isotropic materials, 57
in olivine, 57
Integrated computational materials
engineering (ICME), 105
interfaces
as agents of phase changes, 41
invariance, 52
invisibility cloak, 122
irreversibility, 6
and entropy increase, 6

life, 125
creation of, 131
definition, 125
physics of, 126
synthetic DNA, 131
local isomorphism, 63

magnetic dipole moment, 86
magnetic field
origins, 86
magnetic hard drives, 90
magnetic permeability, 120
magnetisation

definition, 86
magnetism, 86
magnetocaloric effect, 112
magnetoresistance of a ferromagnet, 89
magnetostrictive effect, 112
material
definition, vii
materials
as complex systems, 104
materials design, 102
example, 105
fabrication and service conditions, 104
metamaterials, 114, 123
role of microstructure, 103
self-assembly, 105
systems approach, 104
materials discovery, 103
compared with materials design, 113
materials science
in materials design, 113
and earth sciences, 40, 50, 99, 123
evolution from metallurgy, 46
multiscale character, 89, 95
reductionism, 94
materials selection, 103
measurement
in quantum mechanics, 69
metal-forming processes, 40
metal-insulator transition, 97
metamaterial, 114
contrast with conventional material, 114
elastic, 115
electromagnetic, 118
invisibility cloak, 122
negative refraction, 118
transformation optics, 123
metastability, 1
microstructure, 103
relaxation times, 104
mobility, 35

nanomaterial
chemical reactivity, 83
definition, 82
quantum dot, 83
size dependence of colour, 83
size-dependent properties, 83
surface to volume ratio, 82
nanoscale
definition, 82
nanoscience, 81
nanostructure diagram, 111
nanostructured materials
definition, 82
Neumann's principle, 55
Noether's theorem, 54
converse, 54

olivine, 57
optical response of metals, 120

Ostwald ripening, 110
oxidation, 40

paramagnetism, 88
particle in a box, 83
Pauli exclusion principle, 71
    consequences, 72
Periodic Table, 73
phase, 2
phase diagram, 19
    common tangent construction, 23
    complete miscibility, 25
    eutectic, 26
    intermediate phase, 24
    limited solid solubility, 26
    peritectic, 27
    water-halite, 19
phase rule, 17, 28
phason, 63
phonons, 78
photocatalysis, 112
photochromic effect, 111
photoelectric effect, 67
photomechanical effect, 111
photovoltaic effect, 111
piezoelectric effect, 111
Planck's constant, 55
plasma oscillations, 121
plastic deformation
    and chemical bonding, 48
    occurs by shear, 46
    slip, 46
    slip plane, 46
    stored energy, 51
plasticity, 98
    multi-scale nature, 99
point defects, 41, 45
    Frenkel defect, 45
    impurities, 44
    interstitial impurities, 45
    Schottky defect, 45
    substitutional impurities, 45
    vacancies, 41
    vacancy-impurity complexes, 45
points defects
    in ionic crystals, 45
polycrystal, 41, 50, 53
principle of symmetry compensation, 53

quantum diffusion, 79
quantum dot, 83
    applications, 85
    fluorescence, 85
    light emission, 84
    self-assembly, 108
    synthesis, 86
quantum numbers
    free electron, 55
    hydrogen atom, 72
    particle in a box, 84

quantum theory
    angular momentum, 87
    band theory, 74
    chemistry of the elements, 73
    diffusion, 79
    exchange interaction, 72
    exclusion principle, 71
    harmonic oscillator, 78
    hydrogen atom, 72
    indistinguishable particles, 71
    metals, semiconductors and insulators, 74
    Periodic Table, 73
    probability, 69
    probability amplitude, 69
    quantum state, 72
    solids as giant molecules, 74
    specific heat, 78
    spin, 87
    stability of matter, 76
    transition to Newtonian physics, 70
    tunnelling, 76
    uncertainty principle, 6, 70
    wave function, 73
quasicrystals, 58
quasiparticles, 75
quasiperiodicity, 62

recrystallisation, 51
reductionism, 94, 127
refractive index
    definition, 119
reptation, 37
resonance, 115
reversibility, 7

scanning tunnelling microscope, 76
Schottky defect, 45
screw dislocation, 48
self-assembly, 105
    bubble raft, 106
    photonic crystals, 107
    quantum dots, 108
    vs. self-organisation, 105
self-cleaning glass, 112
self-healing materials, 112
    concrete, 112
self-interstitial, 42
    created by irradiation, 43
self-organisation, 98, 125
self-organised criticality, 99
    in work hardening, 99
shape-memory effect, 111
Simmons and Balluffi experiment, 43
sintering, 41
slip, 46
slip plane, 46
smart materials, 111
Snell's law, 119
specific heat
    Debye model, 78

Einstein model, 78
electronic contribution, 78
spin, 72, 87
majority and minority, 88
spintronics, 90
spontaneous process, 6
stability, 2
state function, 7
stress, 47
superhydrophilicity, 112
surface diffusion, 41
symmetry
broken, 53
compensation principle, 53
continuous, 52
discrete, 52
indistinguishable particles, 71
synthetic biology, 131

theoretical shear strength, 47
thermal activation, 31
thermal expansion, 37
and asymmetry of atomic interactions, 39
and minimum of Gibbs free energy, 39
thermal fluctuations, 31
thermal noise, 36
thermal properties of solids, 76
thermochromic effect, 111
thermodynamic
absolute temperature, 3
chemical potential, 4, 15
components, 2
dynamic equilibrium, 32
entropy, 5
equation of state, 3
equilibrium, 3, 15, 17, 31
extensive variables, 3
function of state, 7
intensive variables, 3
limit, 32
phase, 2

state variables, 3
work, 4
thermodynamic system, 2
closed, 2
isolated, 2
open, 2
thermodynamics
combined first and second laws, 16
first law, 4
second law, 5, 6
thermoelectric effect, 112
topological defects, 57
transformation optics, 123
tunnel magnetoresistance effect, 93
tunnelling, 76
and chemical bonding, 76
scanning tunnelling microscope, 76

uncertainty principle, 6, 70

vacancy, 41
charged in insulators, 44
configurational entropy, 42
dynamic equilibrium, 42
equilibrium concentration, 42
free energy of formation, 42
free energy of migration, 42
measurement of formation free energy, 43
migration energy, 44
valence electrons, 73
magentic moment, 87

wave function, 73
work, 4
chemical, 16

Young's modulus, 55

zero point energy, 39, 70
zero point motion, 32, 70